# SACRED GEOGRAPHIES OF ANCIENT AMAZONIA

# NEW FRONTIERS IN HISTORICAL ECOLOGY

Dynamic new research in the genuinely interdisciplinary field of historical ecology is flourishing in restoration and landscape ecology, geography, forestry and range management, park design, biology, cultural anthropology, and anthropological archaeology. Historical ecology corrects the flaws of previous ecosystems and disequilibrium paradigms by constructing transdisciplinary histories of landscapes and regions that recognize the significance of human activity and the power of all forms of knowledge. The preferred theoretical approach of younger scholars in many social and natural science disciplines, historical ecology is also being put into practice around the world by such organizations as the UNESCO. The series fosters the next generation of scholars offering a sophisticated grasp of human-environmental interrelationships. The series editors invite proposals for cutting edge books that break new ground in theory or in the practical application of the historical ecology paradigm to contemporary problems.

**General Editors**
William Balée, *Tulane University*
Carole L. Crumley, *University of North Carolina, Chapel Hill*

**Editorial Advisory Board**
Wendy Ashmore, *University of California, Riverside*
Peter Brosius, *University of Georgia*
Lyle Campbell, *University of Utah*
Philippe Descola, *Collège de France*
Dave Egan, *Northern Arizona University*
Rebecca Hardin, *University of Michigan*
Edvard Hviding, *University of Bergen*
William Marquardt, *University of Florida*
Kenneth R. Olwig, *Swedish University of Agricultural Sciences*
Gustavo Politis, *Universidad de la Plata*
Nathan Sayre, *University of California, Berkeley*
Stephan Schwartzman, *Environmental Defense Fund*

**Series Titles**
Vol. 1: *Social and Ecological History of the Pyrenees: State, Market, and Landscape*, Ismael Vaccaro and Oriol Beltran, eds.

Vol. 2: *The Ten-Thousand Year Fever: Rethinking Human and Wild Primate Malarias*, Loretta A. Cormier

Vol. 3: *Sacred Geographies of Ancient Amazonia: Historical Ecology of Social Complexity*, Denise P. Schaan

# SACRED GEOGRAPHIES OF ANCIENT AMAZONIA

Historical Ecology of Social Complexity

Denise P. Schaan

LONDON AND NEW YORK

First published 2012 by Left Coast Press, Inc.

Published 2016 by Routledge
2 Park Square, Milton Park, Abingdon, Oxon OX14 4RN
711 Third Avenue, New York, NY 10017, USA

*Routledge is an imprint of the Taylor & Francis Group, an informa business*

Copyright © 2012 Taylor & Francis

All rights reserved. No part of this book may be reprinted or reproduced or utilised in any form or by any electronic, mechanical, or other means, now known or hereafter invented, including photocopying and recording, or in any information storage or retrieval system, without permission in writing from the publishers.

Notice:
Product or corporate names may be trademarks or registered trademarks, and are used only for identification and explanation without intent to infringe.

Library of Congress Cataloging-in-Publication Data:

Schaan, Denise Pahl, 1962-
Sacred geographies of ancient Amazonia : historical ecology of social complexity / Denise P. Schaan.
    p. cm. — (New frontiers in historical ecology; vol. 3)
  Includes bibliographical references and index.
    ISBN 978-1-59874-505-4 (hardcover : alk. paper) — ISBN 978-1-61132-799-1 (ebook)
1. Indians of South America—Amazon River Region—Antiquities. 2. Indigenous peoples—Ecology—Amazon River Region. 3. Human geography—Amazon River Region. 4. Social archaeology—Amazon River Region. 5. Indian pottery—Amazon River Region. 6. Petrolgyphs—Amazon River Region. 7. Rock paintings—Amazon River Region. 8. Amazon River Region—Antiquities. I. Title.
  F2519.1.A6S243 2011
  981'.1—dc23
                    2011021600

Cover design by Jane Burton

ISBN 978-1-59874-505-4 hardcover

# Contents

| | | |
|---|---|---|
| List of Figures and Table | | 7 |
| 1. | Introduction: Historical Ecology and Archaeological Landscapes in Amazonia | 9 |
| 2. | Moving Earth, Managing Water | 29 |
| 3. | Land of the Ancestors | 79 |
| 4. | Ponds, Lakes, and Feasts: The Cultural Geography of Anthropogenic Soils | 105 |
| 5. | Marks on the Earth: Territoriality and Memory | 141 |
| 6. | Conclusion | 177 |
| References | | 195 |
| Index | | 221 |
| About the Author | | 233 |

# List of Figures and Table

Figures

| | | |
|---|---|---|
| Figure 1.1 | Map of Amazonia | 12 |
| Figure 1.2 | Painted rock walls in Monte Alegre | 13 |
| Figure 1.3 | State of Amapá archaeological findings | 23 |
| Figure 1.4 | Probable distribution of Arawak and other language groups | 25 |
| Figure 2.1 | Marajó Island environments | 31 |
| Figure 2.2 | Marajó Island cultural sequence | 34 |
| Figure 2.3 | Eastern Marajó Island archaeological sites | 35 |
| Figure 2.4 | Upper Goiapi River sites | 41 |
| Figure 2.5 | Excavation of an anthropomorphic burial vessel at the Casa Velha 2 site | 43 |
| Figure 2.6 | Distribution of mounds along the Camutins River | 46 |
| Figure 2.7 | Distribution of mounds | 54 |
| Figure 2.8 | East-west cross-section of M-1 | 56 |
| Figure 2.9 | Clay floor in M-1 Excavation 2 | 58 |
| Figure 2.10 | Features at M-1 | 58 |
| Figure 2.11 | M-17 plan topographic map | 61 |
| Figure 2.12 | M-17 features | 62 |
| Figure 3.1 | Burials at M-17 mound | 85 |
| Figure 3.2 | Realistic representations on Marajoara pottery | 87 |
| Figure 3.3 | Pottery decorated with geometric designs | 88 |
| Figure 3.4 | Representations of snakes and/or snakeskin patterns in Marajoara art | 89 |
| Figure 3.5 | Joanes painted funerary vessels | 93 |

| Figure 3.6 | Pacoval Incised funerary urns | 95 |
| Figure 3.7 | Tanga iconography | 99 |
| Figure 3.8 | Figurines: Styles and patterns of breakage | 103 |
| Figure 4.1 | Archaeological sites containing Incised and Punctate Tradition ceramic styles | 114 |
| Figure 4.2 | Incised and Punctate Tradition ceramics | 116 |
| Figure 4.3 | (A) Excavation of a *bolsão* at the Port site, Santarém; (B) Terra Preta do Jacú site, Belterra Plateau; (C) Serra do Capiranga site; (D) pond excavation at the Zinha site | 118 |
| Figure 4.4 | Stone *muiraquitãs* from the Museu do Forte collection | 127 |
| Figure 4.5 | Stone idols, Óbidos | 129 |
| Figure 4.6 | Tapajó pottery | 133 |
| Figure 5.1 | Face vessel | 143 |
| Figure 5.2 | Seu Chiquinho and Colorada Ranch | 144 |
| Figure 5.3 | Location of the ditched enclosures (geoglyphs) in the upper Purus basin | 146 |
| Figure 5.4 | Examples of different site layouts | 148 |
| Figure 5.5 | Spatial distribution of site shapes | 151 |
| Figure 5.6 | Fazenda Iquiri site: Mounds and ditched enclosure | 153 |
| Figure 5.7 | Regional distribution of sites | 154 |
| Figure 5.8 | Fazenda Colorada plan | 156 |
| Figure 5.9 | Fazenda Atlântica vessel | 158 |
| Figure 5.10 | Beni raised fields, Bolivian Amazonia | 164 |
| Figure 5.11 | Map of the western Amazonia showing the locations of indigenous groups | 174 |

Table

| Table 5.1 | Radiocarbon dates | 152 |

*Chapter 1*

# INTRODUCTION: HISTORICAL ECOLOGY AND ARCHAEOLOGICAL LANDSCAPES IN AMAZONIA

ARCHAEOLOGY, AS HISTORY, offers a long-term perspective on society and culture. However, archaeology is the only discipline capable of recording the history of the societies that populated the pre-Columbian Americas during the several millennia that preceded the European conquest of the continent; with the exception of the Mayan Chilam Balam books, the original Americans did not leave any other written accounts.

This book is about the indigenous history of Amazonia—a history that has been written by archaeologists, geographers, ethnobotanists, soil scientists, ethnologists, and geologists—so it draws from diverse fields of inquiry and multiple lines of evidence. As I narrate this history I will define clearly both the information sources I used and the paths I followed to build up this complex, beautiful, and yet incomplete mosaic of life in the Amazonian tropical lowlands.

Human life in the moist South American tropics has been depicted frequently as a model of adaptation. The image of savage tribes of unclothed women, men, and children on the edge of starvation, surviving in the mysterious and dangerous jungle that covers a significant portion of most northern South American countries, is still part of the western imagination. This picture has been formed by more than 400 years of confrontations between indigenous societies and intruders into their lands. This image, however, is flawed, as it does not take into consideration the history of the resistance and the struggle of native peoples to maintain their cultures and defend their territories (Hemming 1978). It does not portray a way of life that is a culmination of a profound knowledge of plant and animal dynamics, crop domestication, hydraulic management, soil mulching, earth moving, the invention of ceramics, and other achievements that I will describe throughout the following chapters. A new approach is in order, one that rescues the natives' point of view.

The historical ecology perspective allows for a portrait of indigenous history that is both fair and in line with the available scientific evidence (Balée 1998, 2006; Crumley 1994). Historical ecology takes into account the historicity of living places, not decoupling the environment from the people who live and have lived in it, but looking at society and the environment through a dialectic perspective. Indeed, it might sound obvious to say that people transform their surroundings, taking advantage of natural resources to satisfy their more diverse needs for food, shelter, and leisure. We know now from firsthand experience that overexploitation of the environment takes its toll, resulting in polluted waters, smoky air, and infertile soils, not to mention climate change and related environmental problems. The extent of human interaction with the environment in pre-Columbian times is less clear, however, and many of us still cling to an image of naked savages living in an immemorial, timeless Eden.

We know that our species, *Homo sapiens*, differs from other primates predominantly in our ability to create culture—as symbolic action—and transform our surroundings; why then do we deny the first peoples of Amazonia the same credit? This just does not make sense. To different degrees, and with different consequences, all human beings transform their surroundings (Balée 2006; Redman 1999; Sauer 1969) in their struggle to build a life—supposedly better—for themselves and their offspring. This does not happen as a single episode, but over many generations of accumulated know-how on managing landscapes.

Historical ecology is thus a perspective that opens a window onto the long-term relationship between people and their environment. Historical geographers and historical ecologists have taught us that landscape is a result of the meeting between society and the environment (Balée 2006; Balée and Erickson 2006; Crumley 1994; Sauer 1969). Within this framework, landscape is not a beautiful painting of bucolic scenery to be hung in the living room, but rather an ebullient place or set of places pregnant with history. As a layered cake, landscape is shaped by the unfolding of countless episodes of building, destruction, and the rebuilding of places and living beings. Landscape is thus the historical ecology unit of analysis.

The study of landscapes, which by definition are anthropogenic and in constant change, requires a multidimensional analysis in regard to space and time (Crumley 1994). Scale is important for the understanding of features and patterns. Different scales lead to different readings of the reality. The landscape approach requires questioning current notions on what a site or a region is, and assumes boundaries as permeable, elastic, and dual, depending on the researcher and society's perspectives (Crumley and Marquardt 1990; Dunnell 1992). Although intentionality is important to understand how landscapes are built (Balée and Erickson 2006), human choices very often affect landscapes in unintended and unpredicted ways, requiring changes in decision-making. Thus, landscapes also reflect conflicts and contradictions within human societies (Crumley and Marquardt 1990).

In the remainder of this introductory chapter, I will present a rapid overview of the history of societies and landscape transformation in the Amazon region in the last 11,000 years. Despite the fact that a lot of research remains to be done and that there are many areas and periods about which we have no information, organizing what we do know through a historical ecology perspective allows us to appreciate the living history of peoples and places, thus reconciling society and nature the way Native Americans have taught us to do.

## The First Settlers

Humans first arrived in the Americas at least 16,000 years ago, if we accept—and I think it is reasonable to do so—the oldest dates obtained by archaeologists for the occupation of the Meadowcroft Rockshelter in western Pennsylvania, as well as the meeting of a few people around a fire in Monte Verde, Chile, at almost the same time (Adovasio et al. 1990; Dillehay and Meltzer 1991). It is still difficult, however, to account for the initial process of migration throughout the continent, since, by 11,000 years BP, groups of hunter-gatherers (the so-called Paleoindians) were occupying rock shelters and open-air sites throughout the Americas, from north to south.

The standard model for the occupation of North America after the last glacial period became known as the Clovis First hypothesis (Meltzer 1988). Clovis is a town in New Mexico near which a group of hunters got together between 11,600 and 11,000 years ago, leaving behind projectile points that were named after that locality. Beginning in the 1930s, several Clovis sites were found in North America, leading archaeologists to think that those were the first inhabitants of the territory and that they apparently specialized in hunting big game, wandering around for at least 400 years until their projectile point technology was replaced. A similar way of looking at those first explorers was imported to South America, where scholars believed that the first humans there were specialized big game hunter-gatherers who preferred to live in open environments.

Before the first Paleoindian sites were studied in Amazonia in the 1980s, researchers thought that humans did not penetrate the rainforest until much later. They believed that only after the introduction of agriculture, at a much later date, did fully agricultural human groups begin to colonize the jungle (Bailey and Headland 1991). The idea was that the rainforest would not attract hunters who specialized in large game, which were more likely to exist in open areas. Although large animals, nowadays extinct, did exist in Amazonia (Ranzi 2008) in the late Pleistocene period, there is no evidence that humans ever hunted them. Further research would show that humans were present in the region since at least 11,200 years BP, gathering a variety of fruits and seeds available in the rainforest, hunting small animals near their camp-sites, and taking advantage of the widely available riverine resources (Roosevelt et al. 1996).

The best account of Paleoindian habitation in Amazonia was provided by environmental archaeology research conducted at Caverna da Pedra Pintada (Figure 1.1.), a sandstone cave located along a chain of reefs that overlooks the lower Amazon River margins, in the municipality of Monte Alegre (Roosevelt et al. 1996). The people who periodically visited and inhabited Pedra Pintada between 11,200 and 9,800 years BP produced and used lithic artifacts such as triangular, stemmed bifacial points and other tools, and lived mostly by gathering wild foods from the nearby forest and floodplain (Roosevelt et al. 1996:377). The cave itself, along with dozens of meters of rock walls that embrace the middle section of the Monte Alegre cliff, was painted with red, yellow, and white motifs, depicting caricatured humans (including pregnant women), handprints, animals, geometric figures, and celestial bodies illustrating a number of astronomical phenomena (Davis 2009; Pereira 2003) (Figure 1.2). These paintings were probably messages motivated by the need to delineate and defend a territory.

Data are more profuse for the end of the Paleoindian (11,000–8,500 years BP) and the Archaic (8,500–5,000 years BP) periods, when

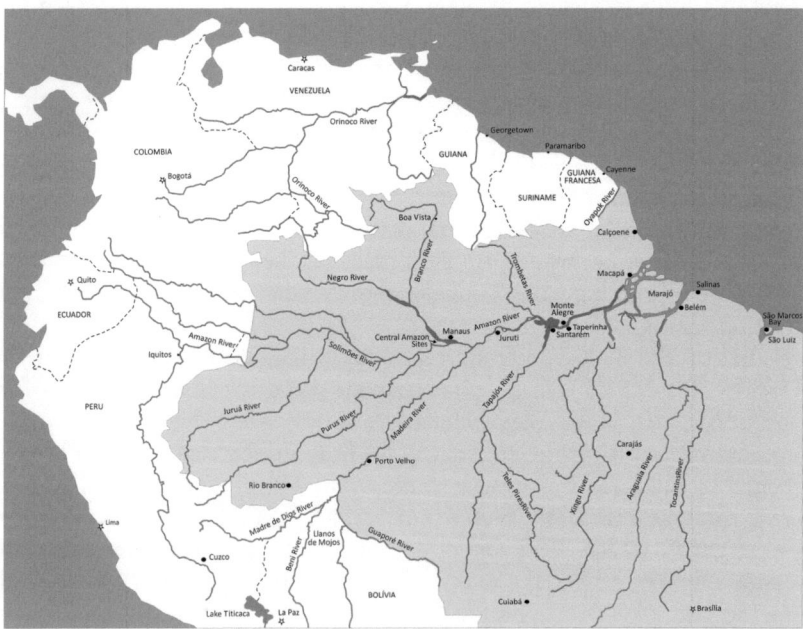

Figure 1.1. Map of Amazonia, showing the places discussed in this chapter (map by Denise Schaan).

Figure 1.2. Painted rock walls at various sites in Monte Alegre, lower Amazon (photographs by Denise Milan).

hunter-gatherers spread over the territory along caves, rock shelters, and riverine neighborhoods. In the Pequiá cave (9,000 years BP), the Gavião cave (6,900 years BP; Magalhães 2005; Silveira 1994), and other sites (10,000–9,000 years BP) in the Carajás mine province, the first Amazonians were living off seeds, nuts, riverine shellfish, and small game, producing myriad artifacts with raw materials that at times originated from far away, such as chalcedony, quartz, silex, and sandstone (Kipnis et al. 2005:87; Roosevelt et al. 1996). Sites were occupied by returning populations for short periods of time. This unspecified, expedient, and abundant lithic industry would survive until 8,000 years BP (Caldarelli et al. 2005), when ceramics came into the picture. It is likely that lithic artifacts were partially replaced by a wood and bone industry, which decayed over time from humidity and did not survive to the present.

Although most hunter-gatherer populations do not transform their surroundings in a very visible way because they are small in numbers and do not practice agriculture, it is possible that they were already building landscapes by managing plants—unconsciously or not—along their hunting paths, creating new forest arrangements (Politis 2001; Roosevelt 2000). This is confirmed by molecular genetic studies of Amazonian crops that indicate an early Holocene domestication of the peach palm, for example (Clement et al. 2010). As studies in this field advance, it is possible that in the coming years, we will have a better knowledge of the full array of forest crops that were managed by forager populations.

## Moving Down to the Rivers and the Sea

In many parts of the world, from Japan (Okada 1998) to Peru (Sandweiss et al. 1998), the first societies to enjoy a sedentary or semi-sedentary way of life relied on aquatic resources. It was not different in Amazonia, a region known for its bountiful rivers and a 932-mile-long stretch of shoreline cut by bays, deltas, and straits, where we find one of the most extensive mangrove areas in the planet. This Atlantic landscape—which begins at the mouth of the Oyapok River (which separates Brazil from French Guiana) and the São Marcos Bay, in Maranhão, marking the eastern limits of the Amazon rainforest—was settled by humans at least 5,500 years ago, thanks to the rich biodiversity available in the region year-round. There, native communities would find shellfish, crab, fish of various sizes, and small game, as well as forest fruits, nuts, and seeds in exceptional abundance. The daily consumption of shellfish produced huge amounts of discarded shell in the habitation sites, which is why such sites are called "shell middens," whereas in Brazil, archaeologists have used the Tupian word *sambaqui*. Why didn't they find the sea earlier? Well, they may have. It is possible that many earlier sites are now underwater due to the sinking of the continental shelf during the Holocene (Richardson III 1998; Silveira and Schaan 2005). A half-submerged site recently found in Salinas, a beach town located three hours from Belém, the state's capital, reinforces this hypothesis. Moreover, shell midden populations were also found in riverine settings at the lower Amazon (Taperinha site), which indicates an aquatic-based sedentary way of life by 7,000 years BP (Roosevelt et al. 1991), long before the spread of crop cultivation.

Unfortunately, our ability to study those first riverine/maritime-adapted peoples is hampered by the poor state of preservation and looting at most of the important sites belonging to this period. Shell midden sites were mined for the lime industry at the turn of the nineteenth century and have become virtually flattened after decades of exploitation.

In Taperinha, however—a site located in a farm at the lower Amazon River first described by Hartt (1885), a nineteenth-century geologist—a six-meter-deep shell mound deposit excavated during the 1980s by a team led by Anna Roosevelt provided evidence of a way of life based on freshwater shellfish and other faunal species, and also pointed to the production of lithic artifacts and ceramics at an early time (Roosevelt et al. 1991:1622). The dates obtained for the use of baked clay in the production of cooking vessels are the oldest for the entire continent, challenging the prevalent theory at the time, which accounted for the arrival of ceramics in the New World by transpacific diffusion from Japan (Estrada 1961; Meggers et al. 1965). According to this diffusionist model, the technology for the production of Middle Jomon–period ceramics (6,000–4,000 years BP) would have arrived at the northern South American coast by 4,000 years BP, where it was readily adopted, and then spread throughout the continent. This hypothesis was based both on the similarities between the two ceramic industries and on the assumption that ceramic technology was invented only once, later travelling around the globe (Meggers 1998).

Taperinha's well-preserved deposits, however, contained ceramic sherds 3,000 years earlier than Ecuador, proving that ceramics were once produced—for the first time in the Americas—in the heart of the Amazon rainforest, although ceramics did not constitute the bulk of cultural vestiges for that period, as it would in the millennia to come. Taperinha's deposits also yielded lithic implements and utensils, such as hammerstones and flake tools as well as grinding and cooking stones. Turtles and shellfish, widely used in the diet, also provided shells that were used as raw materials for domestic utensils (Roosevelt et al. 1991).

Among the Archaic period (from 8,000 years BP until the adoption of farming, by ca. 4,000 years BP, also known as the Formative period) deposits at the site of Caverna da Pedra Pintada, which were separated from the Paleoindian period strata by a 30-centimeter-thick sterile layer, Roosevelt and her colleagues found remains of turtles, fish, shell beads, burned seeds, and ceramic sherds. The pottery, tempered with shells and sand and decorated with incisions and punctuation, belonged to the period between 7,580 and 6,625 years BP (Roosevelt et al. 1996).

In the northern Atlantic coast, where aquatic fauna was obtained from the sea and neighboring mangroves, shell-tempered ceramics were produced between 5,500 and 3,000 years BP (Roosevelt 1995; Simões 1981). In one of the shell middens studied in Maranhão, ceramic vessels surprisingly were found in a layer below shellfish remains, suggesting that ceramics were used before mollusks were incorporated into the local diet (Bandeira 2008). Unfortunately, there is no available chronology for those layers.

As in other parts of Brazil, a diet based on aquatic resources, attested mainly by the presence of prominent shell mounds, continued well into the Formative period. In the lower Xingu River, sites containing shells, animal bones, and shell-tempered ceramics were dated to $3{,}170 \pm 120$ and $1{,}650 \pm 70$ years BP. In two other sites in the vicinities, mollusks were eaten as part of a diet that included game, fish, and supposedly cultivated manioc (from $2{,}255 \pm 55$ to $840 \pm 60$ years BP; Perota 1992:215).

In the Amazon estuary, where the world's largest fluvial-maritime archipelago is located, shell middens were identified along small rivers situated in the forested portion of Marajó Island, and also at the center and south of the archipelago. There was a predominance of *uruá*, a riverine shellfish, as well as shell-tempered ceramics. Local populations have reported the presence of stone adornments and clusters of funerary ceramic vessels containing human bones, which have yet to be investigated by archaeologists (Schaan et al. 2009b).

Although information on early aquatic-focused settlements is scanty, the distribution of these sites is noteworthy, as they stretch from the eastern Atlantic shore of Amazonia to the lower Amazon region, not far from the mouth of the Tapajós River. This provides evidence of the first attempts to settle permanently along riverine landscapes, a trend that would be fully realized at the end of the first millennium.

## The Development of Agriculture

Archaeological interpretation often relies on indirect evidence. Regarding most parts of pre-Columbian, lowland South America, archaeologists have assumed that plants were cultivated and eaten based on the ceramic remains of pots supposedly used for preparing and cooking foodstuffs. Moreover, since most of the calories ingested today by indigenous societies, riverine populations, peasants, and African descendant communities come from bitter manioc products, it has often been assumed that manioc was the main source of starch far back into the past.

After harvest, manioc roots go through a simple but time-consuming process before they are ready for eating in the form of granulated flour (*farinha*, present in every Amazonian meal), tapioca, and bread (*beiju*). The entire plant is used: the squeezed and cooked manioc juice (*tucupi*) is an ingredient in many dishes; manioc leaves are soaked in water and boiled for several days to remove the hydrogen cyanide they contain, and once cooked, the leaves are ground and prepared with dried meats, sausage, smoked pork, and other meats (*maniçoba*; Carneiro 1983; Heckenberger 1998; Smith 2002).

Eyewitness accounts by European explorers and missionaries point to manioc as a staple food among the Amazon floodplain peoples; smaller proportions of maize were also consumed (Carvajal 1934; de Acuña 1859; Heriarte 1874 [1662]). Archaeological artifacts seem to confirm these accounts. Large, flat ceramic plates found as sherds in archaeological sites are believed to have been used as manioc graters, similar in function to today's iron toasters, which are used to produce the desirable *farinha* (granulated flour). Archaeologically speaking, the problem is that manioc roots do not survive in the moist Amazonian soils as macrobotanical remains to be collected by archaeologists, and few attempts have been made to collect microbotanical remains for phytolith analysis (but see DeBoer 1975; Morcote-Rios 2006; Scheel-Ybert et al. 2010).

Nevertheless, the archaeological record contains evidence of the presence of both manioc and sweet potatoes between 8,000 and 6,000 years BP.

Recently, a genetic study of Amazonian domesticated crops has confirmed that manioc was domesticated around 8,000 years BP, indicating also that other crops—such as peach palm (*Bactris gasipaes*), hot peppers (*Capsicum* spp.), pineapple, ingá (*Inga edulis*), and guaraná (*Paullinia cupana*)—were domesticated very early as well (Clement et al. 2010). The general data point to domestication centers in the Amazonian periphery, where open areas of transitional ecotones (like *campo*-savannas, for example) would favor the cultivation of manioc. These findings agree with the linguistic evidence for the origin of the word manioc—from the Tupi *maniot*. Amazonian scholars have assumed that the Tupi were responsible for the domestication and spread of manioc along with other cultigens. Easily cultivated everywhere, manioc crops adapted to different ecological settings as their carriers—who colonized the Brazilian Atlantic coast from north to south—spread over the hinterland, following the Paraguay and Paraná basins (Brochado 1984; Noelli 1996).

Thus, it seems reasonable to assume that manioc was the main source of carbohydrates for early farming communities, who might have also planted sweet potatoes, beans, squash, and pineapples in addition to managing a variety of palms. Protein would come from small game and fish, and occasionally medium to large animals such as capybaras (*Hyrochoerus capybara*), manatees (*Trichechus inunguis*), and pirarucu (*Arapaima gigas*), among others. Additional nutrients were obtained from fruits and nuts, which were always abundant in the rainforest.

At this point, people lived in small villages consisting of two to four houses, or even in isolated houses along the rivers. The population of each village would range from 40 to 300 people (Myers 1981). These figures are based on the overall size of sites and the amount and types of remains, as well as ethnographic models. Villages were usually located next to sources of spring water and streams, but not always near navigable rivers (Schaan 2004:113).

Small villages following this pattern studied in southern Pará have been interpreted as occupations by Tupi-Guarani groups who lived in that region from the beginning of the first millennium until just before the European conquest (Almeida and Garcia 2008; Silveira et al. 2008). Older dates—between 3,800 and 3,600 years BP—for this poorly studied period have been obtained by Gomes (2005) for sites occupied by horticulturalist potters 100 kilometers south of Santarém, along the left bank of the Tapajós River.

A Formative period occupation was also found at the Pedra Pintada cave, where researchers have found grit-tempered pottery (thick-walled bowls and toasters), carbonized fruit seeds, fish and turtle remains, and five poorly preserved burials (Roosevelt et al. 1996).

On Marajó Island, villages occupied by assumed horticulturalists appear between 3,500 and 3,000 years BP, but there are no data for the period between 2,800 and 2,000 years BP, which is equated with a climate change (i.e., a drier period; Meggers and Danon 1988). Hiatuses in the archaeological records were also observed by Gomes (2006) in the Tapajó region between 3,200 and 2,700 years BP and between 2,500 and 1,800 years BP.

## The Ascent of Chiefs

In the beginning of the Common Era, population increased throughout the basin, which can be observed by the number and size of archaeological sites. Site occupation is denser and continuous, and the amount of ceramics increases, along with the appearance of more complex vessel shapes and decoration. A dark fertile soil is found in those sites, also known as *terra preta* or Amazonian Dark Earth (ADE), frequently associated with mounds, interments, hearths, garbage pits, and other cultural features.

Studies undertaken at *terra preta* sites in Amazonia have revealed that these soils contain unusually high amounts of chemical elements such as phosphoros, calcium, magnesium, zinc, and manganese when compared to original soils, and also display an elevated pH and organic content, which makes them especially fertile (Kern and Kampf 1989; Kern et al. 1999; McCann et al. 2000). These chemical elements were probably added to the soil through the decomposition of organic matter related to human activities (Eidt 1985; Kern et al. 1999; McCann et al. 2000; Woods and McCann 1999). High levels of phosphoros, calcium, and magnesium in ADEs are likely due to animal residue (bones, blood, shells, feces, etc.), whereas vegetal residues (palm roofs, wooden walls, hammocks, basketry, and other organic materials) would be responsible for the presence of elements such as zinc and manganese (Kern 1996).

ADE seems to be, at the beginning, an unintentional product, a side effect of population growth (more people discarding more organic matter), although ADE formation has not been identified in some Formative period sites where high amounts of ceramics indicate fairly high populations (Schaan et al. 2009a). For this reason, it is likely that ADE formation indicates a change in subsistence patterns. I believe that an increasing reliance on fish (massively captured in fish ponds), along with a greater production of dry fish and fish flour for storage, would add more phosphors and organic content to the soil. In any case, vast extensions of ADE also relate to discard patterns. At sites where people continuously clean plazas and houses, there is no accumulation of refuse, and remains of feasting and ceremonies are swept away from plazas and pathways, accumulating in garbage areas. So, even though these places indicate a constant and important concentration of activities, ADE would accumulate in peripheral areas (Schaan et al. 2009a). ADE distribution patterns therefore relate to settlement spatial organization, discard patterns, and the history of occupation of a particular landscape (Erickson 2004).

Along the river bluffs and plateaus that border the Amazon River and some of its main tributaries, such as the Negro, Trombetas, and Tapajós Rivers, huge ADE sites have archaeological records of occupation by regionally organized societies that developed from AD 500 onwards (Neves 2008). These are the *várzea* (Amazonian floodplain) complex societies, described as provinces by sixteenth- and seventeenth-century chroniclers (Carvajal 1934; de Acuña 1859), who observed them fully developed. These stratified societies were reportedly ruled by regional chiefs whose symbolic and political power at times extended for many kilometers, as was the case with the Tapajó province, whose main center was located in today's city of Santarém.

Complex regional polities with varying sociopolitical organizations were spread along regions as diverse as the flooded savannas of Marajó Island and the Llanos de Mojos, in opposite extremities of the basin. These large polities also exerted some influence over populations that lived as far as 75 kilometers from their centers of power, but whose material culture emulated the core's symbols of power (Gomes 2005; Schaan and Silva 2004; Stenborg et al. n.d.). Amazonian regional polities maintained supraregional trade networks that integrated markets through the exchange of raw materials and manufactured products, such as stone axes and stone beads (Lathrap 1973); among them, the highly valued greenstone pendants were believed to be used as leverage in the exchange of wives (Boomert 1987).

The first regional polities in the basin appeared on Marajó Island, along with techniques for the management of lakes and rivers—such as the construction of dams and fishponds—to maximize fish capture in the savannas, where periodical floods and a unique combination of fauna and flora were favorable for fish reproduction. These societies used the dirt extracted during the construction of fishponds to build adjacent mounds up to 12 meters high and 3 hectares in area for the habitation of those who controlled the hydraulic systems. In these places, the elite buried their dead in beautiful funerary vessels, decorated with painted, excised, and incised kinship insignia, producing one the most beautiful ceramic traditions in the Americas (Roosevelt 1991; Schaan 2001c).

Marajoara polities were small communities with 2,000 to 3,000 inhabitants with economies based mainly on the chiefs' and shamans' ability to guarantee plentiful food and access to trade networks with other Amazonian polities. Surviving for up to 900 years, the Marajoara cultural development likely influenced other societies in the region (Schaan 2004). An example would be the funerary vessels found at the Caxiuanã Bay, south of Marajó Island, where excised and incised vessel decorations resemble Marajoara phase techniques and motifs. One of the burials was dated to AD $1,050 \pm 50$ (Coirolo 2007).

Around AD 900, regional polities emerged in central Amazonia, where large ADE sites are associated with lakes and impressive amounts of decorated ceramics eventually used in the construction of small mounds, which resulted from termination rituals. Underneath those deposits, archaeologists have found burials, either inside ceramic vessels or directly in the ground (Machado 2005). These mounds are formed by ADE and huge quantities of decorated broken vessels, as well as other objects produced for feasts. Machado (2005) believes that the mounds are monuments related to the funerary practices of a particular social group, built as markers of important changes in the political or religious power.

In central Amazonia, the emergence of regional societies is associated with ceramics belonging to the Polychrome Tradition, which is expressed locally in the Guarita phase, with its anthropomorphic funerary vessels. However, social complexity seems to appear even earlier, with the Paredão phase and the construction of circular villages and mounds (Morais 2006). The fact that ADE was present even earlier seems to point to a long-term process of population increase and social complexity already in the early first millennium. For example, archaeological surveys conducted both at *várzea* and *terra firme* (interfluve) areas in the region between the confluence of the Negro and Solimões Rivers have indicated the existence of occupations with two sites per square kilometer since 2,300 years BP (Lima 2003). Lima believes that there was an economic symbiosis among populations located at river bluffs and *terra firme* (see Denevan 1996).

Defensive trenches were found at the Açutuba and Lago Grande sites (Donatti 2003), whereas Morais (2006) identified mounds, habitation features, funerary contexts, and, hypothetically, turtle corrals in sites located around the Limão Lake. Changes in settlement patterns are visible at the end of the first millennium. Rebellato (2007) explains that at the Hatahara site, Paredão phase occupations (between AD 650 and AD 1000), which were characterized by a circular village plan, changed to a lineal village layout during the Guarita phase (AD 1000–1300).

It is possible that all sites in central Amazonia were highly dependent on fish as a protein source, considering that *várzea* lakes are generally very rich in fish. Along the Trombetas River, the development of a sociopolitical complexity also seems to relate to intensive fishing, since extensive ADE sites containing ritual ceramics are found in the high bluffs that surround lakes seasonally connected to the river (Guapindaia 2009).

The Tapajó domain—situated at the mouth of the Tapajós River, in the area occupied by the city of Santarém today—was described by chroniclers and missionaries with more detail than any other in the basin. According to ethnohistorical sources and archaeological research, we know that the Tapajó were socially stratified, had slaves, and operated within an economy based on maize, manioc, and fish. The main settlements, located at Aldeia ("village") and Porto—two large sites in Santarém—have been studied by Roosevelt and myself (Quinn 2004; Roosevelt 1999; Schaan 2010b). The main vestiges are ceramic sherds, various lithic tools, funerary vessels containing ashes and decomposed bones, and ritual caches full of richly decorated ceramics and feasting remains. The incised, punctated ceramics were dated between AD 1290 and AD 1480 (Quinn 2004). Excavations at the Porto site have revealed deposits that span from the early Formative period to the development of complex societies.

Tapajó ceramics are found in sites located at the mouth of the lower Tapajós River and on the river bluffs and adjacent plateaus, where Nimuendajú, a German ethnologist working for the Göteborg Museum, located sites during the early twentieth century. The same pottery style is also found at huge ADE sites in other parts of the Amazon such as Juruti, indicating a great flourishing of complex societies over a vast region a few centuries before the European conquest (Costa 2009; Sousa 2008).

In the last millennium, vestiges of late pre-Columbian societies were also found at the Pedra Pintada cave. These vestiges indicate maize cultivation, the use of calabashes, and the consumption of fruits adapted to the dry season, such as cashews; this seems to indicate a period when open areas spread due to diminished precipitation (Roosevelt et al. 1996). Ceramics tempered with *cauixi* (sponge spicules) found at that site were dated between AD 1000 and AD 1600 and belong to the Incised Punctate Tradition, which is related to the Tapajó period occupation. In the upper levels, *caraipé* (crushed ashes of the siliceous *Licania scabra* bark) and grog-tempered ceramics related to the Polychrome Tradition were dated between 675 and 430 years BP (Roosevelt et al. 1996).

In the upper Xingu River, fortified villages date between AD 1200 and AD 1600, when excavated half-moon ditches and lineal mounds were built around circular villages. Heckenberger has also identified reservoirs, dams, canals, and roads that connected settlements, creating what he calls an "urban" landscape (Heckenberger 2005; Heckenberger et al. 2003; Villas-Boas and Villas-Boas 1985). Earthworks there and in the Bolivian Mojos have been credited to the expansion of the Arawak peoples (Eriksen 2011; Heckenberger 2005).

Since the 1960s, geographers and archaeologists have studied impressive anthropogenic landscapes at Llanos de Mojos, featuring elevated fields for crop cultivation, fish weirs, ponds, and reservoirs (Denevan 1963, 1980; Erickson 2001, 2006). Ring ditches, first identified in Bolivia, have also been found over the last few years in the *terra firme* region between the Acre, Iquiri, and Abunã Rivers, at the upper Purus River. Spread over a vast area in western Amazonia, a number of earthworks consisting of excavated ditches and embankments displaying diverse geometric shapes date to the beginning of the first millennium. In Brazil, they have been called geoglyphs, and have been interpreted as ceremonial and/or gathering sites, given the low density of cultural remains. Some could have been habitation sites, but the available evidence suggests they were meeting places, or ceremonial sites (Pärssinen et al. 2009; Schaan et al. n.d.).

The discovery of large earthwork sites in the Amazonian periphery, especially in *terra firme* environments, has challenged the once-popular *várzea/terra firme* dichotomy, which considered these to be internally homogenous ecological settings that provided contrasting opportunities for cultural development. From the cultural ecology point of view, the floodplains would have supported larger populations because of their more fertile soils (Meggers 1971; Roosevelt 1980; Steward 1948). The archaeological record, however, has revealed that both landscapes were able to support large regional societies.

In Amapá, the Brazilian Guiana, there are vestiges of at least four different cultures beginning in AD 1000 and lasting until the arrival of the Europeans. Sites generally contain urn burials. One type consists of burials under huge blocks of granite in open spaces surrounded by megaliths with possible astronomical significance (Figure 1.3.A–B; Cabral and Saldanha 2007, 2008). Nineteen of those sites, located in the municipality of Calçoene, have been assigned to ancestral Arawak peoples.

In the Maracá River area, funerary anthropomorphic vessels representing women and men sitting on stools are found inside caves (Figure 1.3D), with varying distances to habitation sites (Guapindaia 2001). Conversely, in the Cunani River region, urn burials (Figure 1.3C) are found inside boot-shaped burial chambers dug in the ground. Similar practices are known in Colombia, which suggests some relation between the two regions. The Arawak populations, who occupied the northern South American coast in the sixteenth century, could have disseminated some of these customs (Nunes Filho 2005).

In 1491, Amazonian societies were growing in number and in sociopolitical complexity. In the Amazon Delta, the collapse of the Marajoara polities around AD 1350—for reasons not totally understood—happened at the same time that Arawak groups coming from the north began to penetrate the island (Meggers and Evans 1957:422–424, 554–555). The Arawak diaspora reached the entire Amazonian periphery, parts of the Amazon itself, the Negro, Solimões and Madeira Rivers, and the Antilles, thus dominating a good part of northern South America and the Caribbean on the eve of the European conquest (Heckenberger 2005; Figure 1.4).

The Tupian-speaking groups, whose probable origin was the upper Madeira region in Rondônia, were already established along the eastern

Figure 1.3. State of Amapá archaeological findings: (A) Megaliths, Calçoene; (B) Ceremonial cache inside a funerary pit, Calçoene; (C) Cunani vessel; (D) Maracá funerary cave. Photographs by Mariana Cabral (A–B) and Maurício de Paiva (C–D).

coast of the continent, occupying Brazil from north to south. In the Amazon basin, regional polities were in expansion, sustaining complex trade networks that connected the region in all directions, not only enabling the flow of goods and prestige items, but also promoting the interchange of ideas and techniques.

During the transition to the second millennium, social hierarchies and social stratification in northern Bolivia, the upper Xingu River, and in the central and lower Amazon regions developed as people began to transform landscapes, managing earth to create new subsistence systems based on a managed ecology. The presence of burials and of a whole set of ceremonial practices related to shamanistic rites as well as possible ancestor worship indicates the spread of religious systems associated with the new landscapes,

Figure 1.3. Continued

with special agro-managerial connections to the supernatural world. Among the Amazonian regional polities, power was based both on economics and on the symbolic manipulation of landscapes and subsistence resources.

Displaying sophisticated knowledge, these societies were responsible for the building of the diverse and impressive landscapes met by the Europeans in the sixteenth century. Sociopolitical institutions did not develop in any other part of the lowlands as they did in the Amazon basin. There is no point in arguing whether these societies could have turned into expansionist states such as the Inca Empire, but they were certainly closer to that than to the picture of the idyllic tropical forest tribes painted in the 1950s.

This brief overview of sociocultural development in Amazonia is intended to demonstrate that the emergence of regional, complex societies was part of a long-term process of cultural change that became possible when the Amazonian populations began to engage in activities that led to a large-scale transformation of their surroundings, particularly by managing natural resources in more assertive ways.

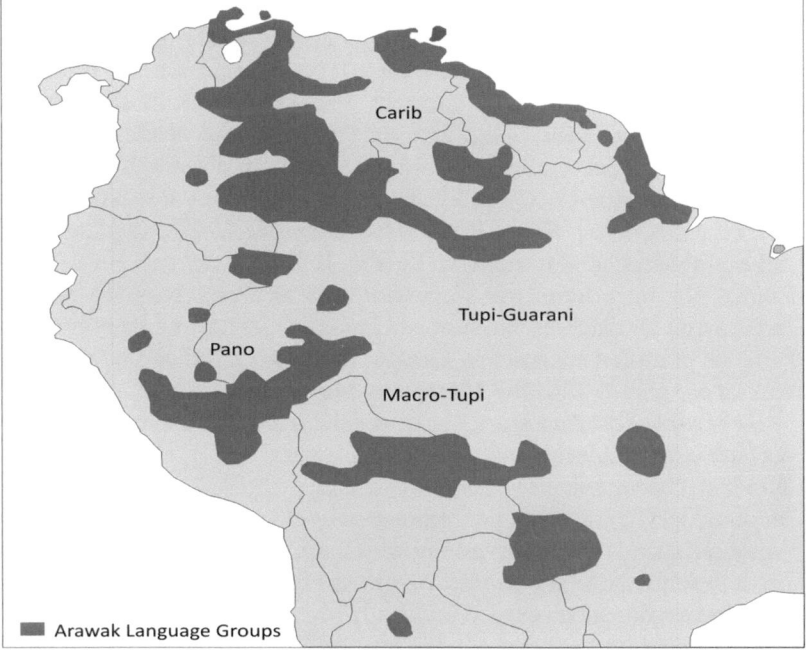

Figure 1.4. Probable distribution of Arawak and other language groups in northern South America at the time of contact. After Aihkenvald (1999).

## THE ORGANIZATION OF THIS BOOK

This book is divided in five main chapters, which focus on the cultural development of three main geographic areas in the last two millennia: the Amazon Delta (Marajó Island savannas), the lower Amazon (Tapajós and Nhamundá/Trombetas Rivers), and western Amazonia (the eastern part of the state of Acre). In these three areas, abundant evidence indicating centuries of landscape management and the development of regional societies with special connections to the supernatural world have led us to examine those societies through a historical ecology perspective.

"Moving Earth, Managing Water" focuses on the archaeology of Marajó Island, where regional societies arose in AD 400. Those societies capitalized on predictable seasonal floods and droughts for the express purpose of managing water and increasing fish capture in dammed rivers and lakes. These earthmovers also built groups of interconnected mounds where they lived, celebrated marriages and births, buried their dead, organized feasts, and performed sacred rituals. Because mound-building has been a recurrent feature in different pre-Columbian societies, I also bring comparative material to bear in the concluding section of the chapter, addressing the particularities of mound-building/sunken plazas at the Faldas de Sangay site, in Ecuadorian Amazonia (Porras 1987; Salazar 2008); the raised-field agriculture in the Guyanas (Rostain 2010); the raised fields for maize cultivation associated with mounds and causeways in the state of Barinas, in the Venezuelan Llanos (Spencer and Redmond 1998; Zucchi 1985); the extensive complexes of ridged fields in Colombia (Parsons and Bowen 1966; Plazas et al. 1993); the defensive earthworks at the upper Xingu River (Heckenberger 2005); and the raised fields and associated mounds in Baures, Bolivia (Erickson 2000, 2006, 2008). The purpose of this discussion is to evaluate both the particularities and similarities between earthworks constructed by contemporaneous complex societies in Greater Amazonia, indicating how they can be compared with the Marajoara phase populations.

In "Land of the Ancestors," I explore the symbolic realm of Marajoara social life, particularly funerary rites, ancestor worship, and the relationship between subsistence practices and the mythological beings depicted on mortuary paraphernalia. Ethnography from various parts of the Amazon basin is incorporated into this chapter. The Tukanoan creation myth that a huge snake-canoe brought the first settlers to earth, placing the sibs (patrilineal clans) hierarchically along the river (Chernela 1982, 1997), finds correspondence in the hierarchical settlement pattern found at the Camutins site on Marajó Island, as well as in the elaborate use of snake iconography in its ceramics.

The ADEs found along the Amazon, Tapajós, and Trombetas Rivers are the focal point of "Ponds, Lakes, and Feasts: The Cultural Geography of Anthropogenic Soils." The Tapajó society extended for about 75 kilometers to the south of its political center, located at the confluence of the Amazon and Tapajós Rivers, along river bluffs and plateaus (Nimuendajú 2004; Schaan 2006). In the uplands, the Tapajós excavated ponds to store rain and spring water. Today, their sites extend over several acres of continuous ADE soils teeming with decorated sherds, parts of ceramic vessels and figurines, stone axes, and spindle whorls (Stenborg et al. n.d.).

Along the Trombetas River, a system of river-connected lakes surrounded by ADE sites rich with Konduri pottery suggests that intensive fishing was important for the rise of complex societies. Besides similarities in settlement patterns and material culture, which have led scholars to assign both societies to the Incised Punctate Tradition (Meggers and Evans 1961), these populations were involved in the greenstone trade network which linked the Caribbean and the Amazon basin on the eve of European contact (Boomert 1987).

In "Marks on the Earth: Territoriality and Memory," I present and discuss the ring ditches, or geoglyphs: large-scale, spectacular enclosures in the form of geometric figures sculpted in the clay soils of western Amazonia. Over an area of roughly 100,000 square kilometers that encompasses the western Brazilian state of Acre and northern Bolivia, these gigantic marks on the ground were revealed to western eyes as deforestation advanced across the region. The 300-meter-wide squares and circles comprising excavated ditches and adjacent earthen walls are a more recent confirmation of the versatility of Amazonian cultures (Pärssinen et al. 2009). Entrenched in an interfluvial region, once thought to be unsuited for permanent settlements, populations living more than a thousand years ago at the Purus River basin were able to organize labor with the purpose of building landscapes full of symbolic significance (Pärssinen et al. 2009; Ranzi 2003; Schaan et al. 2010a). More than 270 earthworks have been discovered so far. This chapter summarizes the most recent data on the configuration and number of sites, combining geographical and archaeological data, and showing the preferred location of these structures in relation to altitude and water resources as well as debates various hypotheses on their possible function and meaning. Possible relations between the Andes and the tropical forest are explored, since the geoglyphs were the midway point between two regions that exchanged finished goods, raw materials, ideas, and language over a long swath of time.

The probable multiple uses for the ditches and walls—defensive, ceremonial, and agricultural—are assessed in all their dimensions. Little mounds found inside and outside of the geoglyphs are compared to similar structures: the Cerritos in southern Brazil and Uruguay (López-Mazz 2001), in central Amazonia (Machado 2005), and the upper Xingu River basin (Heckenberger 2005), where they seem to have been produced by house decay, garbage, and burials.

In the concluding chapter, I summarize the main goal of the book, which is to show that native Amazonians created sacred geographies to obtain resources for their bodies and souls. Bringing the discussion to the present day, I demonstrate that pre-Columbian societies' marks are still unerringly imprinted on the landscape, which is being used, altered, rebuilt, and resignified by contemporary societies still.

I discuss examples of current practices of landscape management among present-day indigenous and peasant populations, which are reminiscent of their indigenous past. I argue that governments tend to disregard traditional, inherited knowledge, imposing instead economic models that are proving to be inappropriate both for the people and the environment. While traditional populations are being pushed out of their lands because of logging, soybean plantations, cattle ranching, mining, and other unsustainable economic enterprises, the tropical environment has endured climate change, a decrease in biodiversity, the disappearance of certain species, desertification, and other human-induced catastrophes. Claims for the preservation of the tropical forests are generally based on mistaken ideas about nature, regarding landscapes that have been managed for native Amazonians for centuries or even millennia as "natural" or pristine. Before we decide whether we want to restore Amazonian landscapes, we need to know exactly what was original about them, and how they were integrated into the vast diversity of phyla and species in the first place. That can only be done if additional research like that presented here can continue into the future.

*Chapter 2*

# Moving Earth, Managing Water

In the beginning of the Common Era, new settlements appeared on the river bluffs along floodplains and lakeshores throughout the Amazon River basin. These indicated an important change in the ways Amazonians would relate to their immediate environment from then on. In places where aquatic resources were more easily harvested, population growth would eventually result in new forms of social organization as well as the emergence of large, regionally organized polities by the end of the first millennium.

During the first Spanish and Portuguese expeditions, chroniclers briefly described some of these large polities in the early Colonial period, emphasizing the variety and abundance of their food resources, their impressive demography, and the existence of broad exchange networks and road systems that linked the riverside settlements to the hinterland. A comparison between chronicles shows inconsistencies regarding the names of the provinces, their demography, and the distances that might be overcome by careful evaluation (Porro 1994); in any case, they are the sole source of eyewitness accounts on what life was like in the Amazon River basin at the onset of the European arrival:

> At once the Captain [Orellana] ordered that he be given clothes and other things, with which he [the Indian Chief] was very pleased...the Captain told him to order nothing to be furnished him but food; and straightway the Chief ordered his Indians to bring food, and in a very short time they brought, in abundance, all that was needed, including meats, partridges, turkeys, and fish of many sorts (Napo River, 1542; Carvajal 1934:175).

> ...the overlord leaped out on land, and with him many important personages and overlords who accompanied him, and he asked permission of the Captain to sit down, and so he seated himself...and he ordered [his followers] to bring from his canoes a great quantity of foodstuffs, not only turtles, but also manatees and other fish, and roasted partridges and cats, and monkeys (upper Amazon; Carvajal 1934:182).

Having arrived at the Machiparo province, located, according to Porro (1996:26) on both Solimões River margins, between the Tefé and Coari Rivers, one of Orellana's men who accompanied him on his overland journey recalled:

> [they] told him exactly how things were and that there was a great quantity of food, such as turtles in pens and pools of water, and a great deal of meat and fish and biscuit, and all this in such great abundance that there was enough to feed an expeditionary force of one thousand men for one year (Carvajal 1934:192).

Carvajal (1934:200) also mentioned that "there were many roads that entered into the interior of the land, very fine highways."

As vague as they might be, ethnohistorical sources about the central and upper Amazon River basin may stimulate hypotheses to be tested in the archaeological record. Unfortunately, however, no accounts were given at that time of the people who inhabited Marajó Island, the largest island in the archipelago located at the mouth of the Amazon River. The first written accounts of Marajoara populations would only be provided by the mid-seventeenth century, when extensive trade, the spread of diseases, slavery, and missionization had already either dispersed or decimated most of the indigenous population. Even so, only the indigenous peoples living on the eastern coast were briefly described. Outsiders never knew the way of life of the populations who lived on huge earthen platforms built over the flooded savannas.

Indeed, it was only in the second half of the nineteenth century that scientists and government personnel began reporting that spectacular pottery vessels and objects were being found in huge mounds spotted across the savannas, testimonies of a vanished culture (Ferreira Penna 1877, 1885; Marajó 1895). Thus, as scholars explored these rich deposits of the Marajoara societies, whose life and death were surrounded by mystery, Amazonian archaeology was born. Therefore, since its beginnings, archaeology has been the only source of information on the Marajoara phase way of life.

## An Island of Contrasts

Marajó is the largest fluvial-maritime island in the world, with an area of 49,606 square kilometers, comparable to the size of Belgium or almost the size of West Virginia. Because it is located at the mouth of the Amazon River and subject to seasonal flooding, some authors eventually compared it with the Amazonian floodplain, calling it *várzea* (Brochado 1980; Roosevelt 1991). In fact, the half-year-long submersion suggests similar hydrologic systems and sedimentation processes for both areas, both of which influence soil fertility. However, these areas differ markedly from each other. The *campos* ("fields," as savannas are called in Marajó Island) are flooded by rainwater during the tropical winter, while the lower Amazon floodplain is invaded by the river overflow

(Sombroek 1966). Besides that, the eastern coast of Marajó Island does not receive the enormous discharge of muddy, nutrient-rich sediments carried by the Amazon River, which is deposited on the western coast of the island and on a belt of "forest islands" that splits the Amazon River into two main channels before it meets the Atlantic Ocean.

Historically, Marajó Island has been described as divided into two broad, distinct regions: the southwestern forest and the northeastern *campos*. This is a reflection of the disparate geologic history of these two areas, but landscapes here are much more diverse (Figure 2.1). Marajó has patches of upland (*terra firme*) forest, floodplain forest, flooded forest, and upland and lowland savanna, as well as mangroves. These ecologic areas are subject to differential rainfall, species diversity, drainage systems, and soil productivity, not to mention exploitation in various ways by human populations.

The island has no true rocks, but a pseudo-sandstone (*grês do Pará*), which was formed by the migration of iron to the surface. *Grês do Pará* is not suitable for making tools, but was used in the early colonial period as a construction material in both mission buildings and fish weirs on the eastern coast (Lopes 1999). The few archaeological stone tools (mostly axes) and other lithics such as beads and rare greenstones found in archaeological contexts are all derived from elsewhere, through long-distance exchange networks.

The focus of this chapter is the northeastern savanna, which is 23,046 square kilometers in size and is bordered by mangroves and patches of floodplain forest along its northern and eastern coasts. Gallery forests

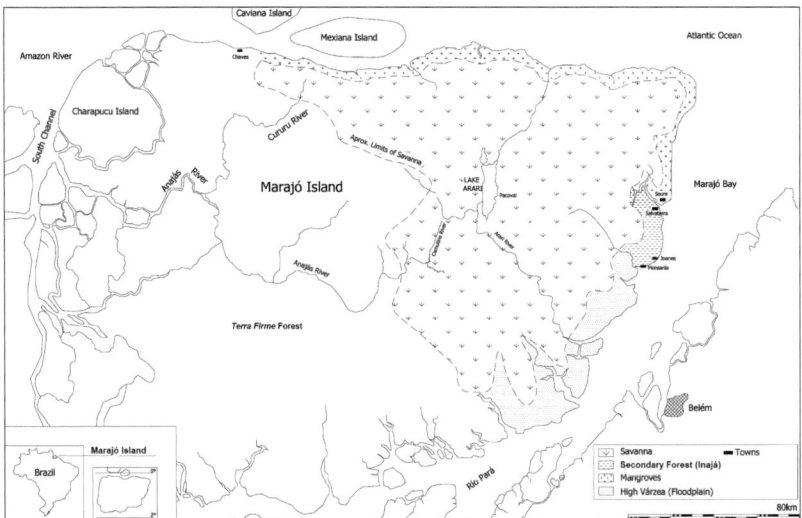

Figure 2.1. Marajó Island environments. After Schaan (2004:48).

frame the rivers, and "forest islands" and patches of shrubs can be found on the predominantly grassy landscape. Palms are also very common in Marajó Island and dominate certain areas, which is a good indication of human management (Balée 1989; Smith 2001). Palms were used by indigenous populations for a wide range of products, from food to craft production, and long-term manipulation of such plant species probably helped define the present landscape.

Rivers in the savannas are ephemeral, tidal rivers, replenished twice daily by the Amazon River tides. Some permanent lakes, such as Lake Arari, are situated roughly in the center of the region, with a major concentration of archaeological sites.

The climate is characterized by two well-defined seasons: a rainy winter (from January to June) and a dry summer (from August to December). Seasonal differences in precipitation lead to important ecological changes in the savannas. The rainfall is intense and continuous for a few months, which causes river overflows and floods, turning the *campos* into large lakes where riverbeds are not discernible.

The highest elevations, no more than 6 meters high, are found on the eastern coast. From there, the land flattens westwards, like a large plate with rising edges (Ackermann 1963:121). During the rainy season, which lasts for up to six months, rainwater accumulates in the deepest center of the "plate." Conversely, during the summer months, the lack of precipitation and the temporary nature of most water sources cause severe desiccation of soils and vegetation. This situation is particularly critical at river headwaters, which depend on occasional rains to maintain a minimal flow. On the other hand, life along the main rivers is less affected because water is available year-round.

The impact of the fluctuation of water on the lowland savannas brings dramatic consequences for human life, for environmental conditions determine the availability of resources as well as different strategies for their exploitation. During the flooded period, transportation across the savannas is possible only by boats and canoes. Pile-dwelling is the preferred method for house construction today, so people are not surprised by the rising waters. During pre-Columbian times, mounds were also built as a means to escape the flood waters (Meggers and Evans 1957; Roosevelt 1991). Fish are scarce during the rainy season because they disperse to spawn and feed on fruits and floating vegetation (Smith 2002). As for protein, hunting is an option, since people and animals are constrained over small patches of land.

When rains cease in July and waters begin to recede, the fish move back to the rivers. Over the savannas, ponds built by pre-Columbian populations retain and concentrate massive amounts of fish. Local populations have learned to build small wood and mud dams across rivers and lakes to capture large quantities of fish. From September to

December, fishing is rewarding at river headwaters all over the savannas. This is the major richness of the island, and it is here that ecology and human management meet to create landscapes that are characterized by resource abundance. The emergence of large regional polities in the area between AD 400 and AD 1350 is related to landscape management and its development over time.

## Antecedents

There is no evidence of a Paleoindian occupation of the island by hunter-gatherers. It seems that the first inhabitants were previously settled populations which relied on shellfish and other aquatic resources. Six shell middens have been registered both in the savannas and the forest, but have not been studied by archaeologists (Schaan and Martins 2010). Shell middens are testimonies to the first societies settled in the Amazon basin, and appear in the lower Amazon at around 7,000 years BP and the Amazonian Atlantic coast around 5,000 years BP (Roosevelt 1995; Roosevelt et al. 1991; Simões 1981). Archaeological research on Marajó Island, however, has focused mainly on the occupation by agricultural peoples and ultimately the chiefdom societies. Therefore, the best-known period is between 3,500 years BP and 750 years BP.

Archaeological research started in the late nineteenth century; by the mid-1940s, many sites had been excavated, some of them to the ground (Hilbert 1952), in search of the beautiful pottery now present in museums worldwide (Schaan 2009b). Theoretically oriented, more comprehensive research was conducted in the late 1940s, when Betty Meggers and Clifford Evans (1957), both Columbia University doctoral students, completed the first regional survey in the north and central areas of Marajó Island. Before them, scholars knew only the extravagant Marajóara ceramic culture.

Meggers and Evans are credited with finding many sites that both predate and postdate the appearance of the Marajoara culture, placing sites in chronological order using pottery seriation based on Ford's (1949) methodology (Figure 2.2). Sites belonging to the period prior to the emergence of complex societies on the island (Marajoara phase) were described as small villages of horticultural peoples living in different parts of the savannas and on the forest borders. The radiocarbon dates place these occupations in two main periods: the Ananatuba and Mangueiras phases (3,460 to 2,800 years BP) and the Formiga phase (2,000 to 1,200 years BP), with a documented hiatus believed to be caused by climatic change, perhaps as a result of survey bias (Meggers and Danon [1988] credit it to climatic change). After the Marajoara culture demise, the Aruan phase is related to people coming from the Caviana and Mexiana Islands, which is interpreted as an invasion of Marajó Island (Meggers and Evans 1957:168–424).

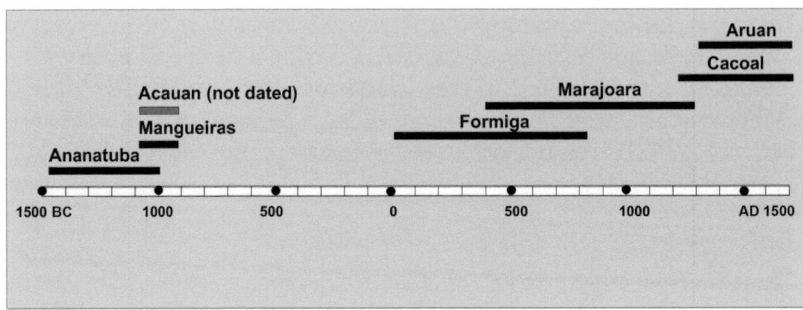

Figure 2.2. Marajó Island cultural sequence.

The cultural sequence presented by Meggers and Evans in the 1950s has not been substantially modified, except for the addition of a late phase contemporary with the Aruan, the Cacoal phase, which can be understood as a post-Marajoara phase (Schaan 1999). Three sites containing different ceramic technology related to this phase were dated between AD 1350 and AD 1600; some of the specimens emulated the Marajoara style. The sites are quite small and are located on the non-flooded banks of the Anajás River, representing autonomous villages.

When the ceramic phases were first described, they were meant to characterize different, unrelated populations that supposedly came to the island from the north, settled there for short periods, and vanished thereafter. Meggers and Evans (1957:421) believed that these cultures were already fully developed by the time they appeared on the island because they had mastered pottery techniques. For the authors, with the exception of the Marajoara phase, all these cultures represented the typical tropical forest pattern, mirrored in ethnographic examples. The tropical forest tribe had been defined as a culture type based on descriptions by ethnographers during the twentieth century (Lowie 1948; Steward 1948). Ethnographic data was simply projected into the past, without considering that major changes had occurred within indigenous societies during the colonial period, with major populations losses and disruption of social and cultural systems. Major differences among the archaeological phases were attributed to their respective ceramic industries. The Ananatuba, Mangueiras, Acauan, and Formiga phases were considered archaeological correlates of different ethnic groups that would have influenced each other's ceramic styles by coexisting in the same geographic area (Figure 2.3).

The Marajoara phase was different, because they built ceremonial (cemetery) and habitation mounds, which implied organization of labor and leadership. An analysis of the cemeteries indicated differential treatment of the dead, compatible with differential ranks or social stratification. Because they appeared to be skillful pottery makers, creating remarkably beautiful and complex ceramics, they probably had

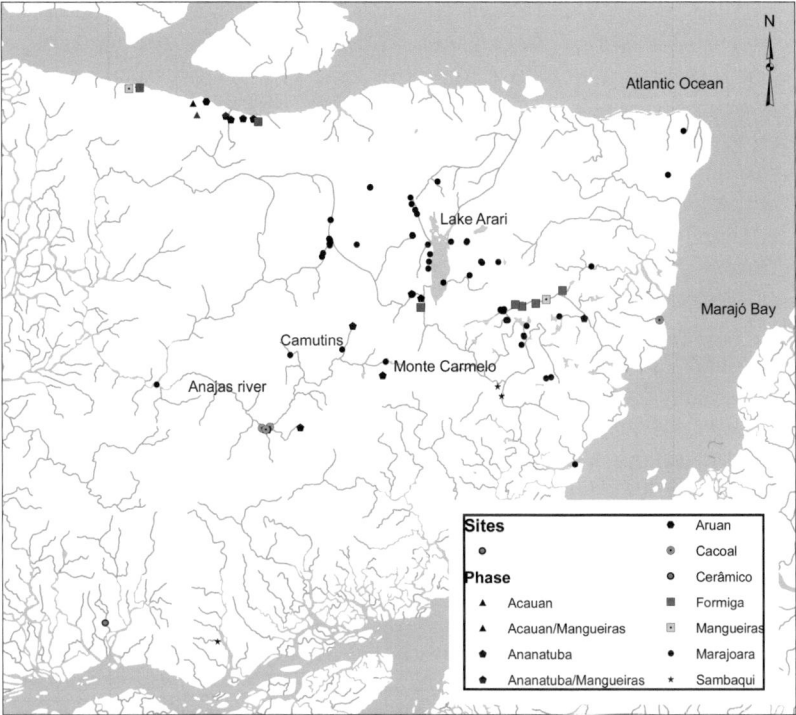

Figure 2.3. Eastern Marajó Island archaeological sites.

specialists, which implied a division of labor found only in complex societies. Pottery objects were used in religious cults and ceremonies, evidence of an elaborate religious system.

Meggers and Evans (1957) did not believe there was any correlation between Marajóara culture and previous cultures, nor did they think that all these abilities (mound-building, sophisticated pottery) were developed through time. On the contrary, some of the Marajoara phase cultural traits, such as secondary urn burials, the use of stools, and hollow rims, had been identified among other cultures in northwestern South America, which, according to Meggers and Evans, provided evidence for migration (Meggers and Evans 1957:Figure 146). Meggers and Evans stated that the Marajoara phase represented an advanced culture of the circum-Caribbean or sub-Andean level of social organization, of a people who arrived on Marajó Island at the peak of their development. This line of reasoning was based on cultural determinism and neo-evolutionism, which held that the immediate environment determined subsistence patterns, technology, social organization, and all other aspects within a culture. Ecology was evaluated independently of culture, and they did not consider human agency as a possibility.

When Meggers and Evans (1957:423–424) built their archaeological sequence for Marajó Island, radiocarbon dating was not available, so they presented a relative chronology based on the presence of ceramics from different phases in the same sites, and on assumptions about the correlation between the thickness of the archaeological deposits and calendar years (10 centimeters = 100 years). When radiocarbon and thermoluminescence dating became available, the order of the sequence remained the same, but occupation times were longer. The Marajoara phase in particular, which they assumed was fully developed at its onset and decayed over time (the first estimate for its duration was 200 years), had survived at least 900 years in what was supposed to be a poor environment. Thus, absolute chronology created a major problem for the ecologically based hypothesis of cultural involution.

Later, Meggers proposed an explanation to what she called "The Mystery of the Marajoara," which claimed an "ecological solution" (Meggers 2001). Puzzled by the long permanence of Marajoara populations in a supposedly limited environment, Meggers suggests that they could have survived on palm starch obtained from the numerous buriti palms (*Mauritia flexuosa*) on the island. Although the consumption of palm starch should not be ruled out, it is unlikely, since its use is unknown among lower Amazon River populations. Moreover, the use of a wild source of starch would not be enough to explain the development of the complex sociopolitical organization of Marajoara culture through time, nor its expansion over an area of 20,000 square kilometers.

Given the many critiques the cultural ecology theory received from scholars in several areas of expertise—especially those showing that human populations have the ability to overcome environmental constraints (Carneiro 1961; Denevan 1963; Lathrap 1970b)—Meggers and Evans' account of the cultural development in the island did not have many adepts. However, their general description of the Marajoara phase as a complex society was overall correct, apart from minor errors related to its origin and mound functionality. They had described two basic mound types: ceremonial and habitation. Later, however, Roosevelt (1991) would show that ceremonial mounds also had abundant domestic features; therefore, they were in fact habitation sites for the elite, who buried their dead in the same space.

Two differences should be pointed out regarding Meggers and Roosevelt's interpretations of mound functions. In recognizing that mounds fell under two categories, Meggers correctly characterized settlements as hierarchical: elite/ceremonial mounds were higher, larger, and fewer in number; habitation mounds were smaller, and largely outnumbered the first ones. Since Roosevelt worked only in two elite/ceremonial mounds, and had knowledge of many more across the savannas, she disregarded Meggers, and proposed that all mounds contained the same features, which caused her to later deny the existence of a hierarchy within multiple mound sites (Roosevelt 1999:21).

Roosevelt can be credited with pointing out that the Marajoara populations had evolved locally, although she does not explain exactly what she means by that, sometimes making it seem that the source of their culture is someplace in the lower Amazon (Roosevelt 1991:3). It is important to highlight, however, that complex societies developed everywhere from simpler and local prior social formations (Carneiro 2007). In the case of Marajó Island, a long-term acquaintance with local ecological conditions would be necessary to master the environment and learn the necessary strategies to develop efficient subsistence systems.

## Continuity and Change

The Marajoara culture was an emblematic example of social complexity in the lower Amazon River, challenging cultural determinist theories pertaining to sociocultural development in the Amazon basin. Replicating fieldwork in Marajó Island, however, proved costly, and it was contingent upon authorization from local scholars in Brazil, most of whom were aligned with Meggers at the time (Meggers 1985; Roosevelt 1991:105–111). Eventually, Donald Lathrap and his students dedicated themselves to the task of reviewing Meggers and Evans' data using new theoretical perspectives, such as Thomas Myers' history of settlement evolution in Marajó Island (1973).

Using Meggers and Evans' massive report, Myers (1973) analyzed site size and location to evaluate how settlements in the island changed through time. He observed a shift in the shape, size, and location of sites. Oval-shaped settlements from the Ananatuba and Mangueiras phases evolved into lineal mounds during the Formiga and Marajoara phases. Settlement size also increased. At the same time, intentional mound building begins during the Formiga phase. Myers tried to correlate this data with population figures and levels of social organization, suggesting that there was a continuum between the Marajoara phase and the previous ones, both in ceramics technology and mound-building practices.

In the Ananatuba phase, mounds were small (less than 1 hectare), consisting of a single elevation, no more than a meter high, probably produced by the accumulation of refuse. The inhabitants preferred to live in the forest, close to the edge of the *campo* and near water, although the presence of a navigable stream did not appear to be important (Meggers and Evans 1957:193). They were good potters, and some pots were decorated with finely incised, sometimes cross-hatched lines inside bounded areas. Others received a coat of red paint or were brushed. The abundant clay was also used for house construction, since during excavations we found residues of wattle-and-daub. They probably used cotton for nets and hammocks, and also enlarged their earlobes for the insertion of pottery discs.

The Mangueiras phase, partially contemporaneous with the Ananatuba phase, built higher mounds that reached 2.6 meters, also resulting from the accumulation of anthropogenic refuse from 0.1 to 4 hectares in size. Sites are located in nonflooded forest, generally close to a navigable river or stream (Meggers and Evans 1957:221).

Mangueiras phase pottery is similar to that of the earlier Ananatuba phase, although it resembles Marajoara phase pottery. Decorated vessels were abundant, but surface treatment consisted mostly of scraping and brushing, with a few incised potsherds that seemed to emulate Ananatuba phase decoration (Meggers and Evans 1957:215). Although the Mangueiras phase occupation was first found and dated above Ananatuba deposits in the Castanheira site, it was found intermixed with or below Marajoara deposits at other sites, which suggests that the Mangueiras phase was a later occupation (Schaan 2001b).

A number of questions about the Mangueiras occupation remain unanswered. Variability in pottery decoration among sites, as well as inconsistencies in the relative chronology used in the 1960s, seems to indicate that geographic isolation may have been an important cause for these differences rather than chronology. Apart from potsherds, several other ceramic objects were found at the Croari site: a figurine head, a figurine body, a miniature jar, a complete and a fragmented tubular pipe, and a short cylindrical object that appears to be a labret. Objects recovered at the Caviana site include a labret, a broken pipe stem, and an unidentified biconical small object (Meggers and Evans 1957:197–198).

The most intriguing phase is the Acauan, although it is not always mentioned in the sequence because their elaborate ceramics came from a private collection and two poorly preserved sites on Mexiana and Marajó Islands. In Marajó, Acauan sherds were found on the bed of the Jurupucu River, which runs parallel to the northern coast, but inland. Meggers and Evans explained that the site was probably destroyed by the river.

The Acauan phase material showed similarities to the Marajoara phase ceramics both in techniques and motifs. The excised ceramics, in particular, is strikingly similar in many details to those of the Marajoara phase, although Acauan phase ceramics lacks the slips found in other Marajoara excised types. When determining an ancestor to the Marajoara phase, Acauan phase is a good candidate, since it is possible to visualize some evolution in decorative techniques. Vessel forms constitute another similarity between Acauan and Marajoara phase assemblages. Besides potsherds, Meggers and Evans (1957:434–438) collected anthropomorphic and zoomorphic ornaments, two figurine heads, a pottery stamp, and a globular spindle whorl during excavations and surface collections.

The Acauan phase was not dated by absolute methods, but since the sherds that suggested contact were found in "early" Mangueiras sites (assuming the seriation based on plain sherds is correct), Acauan phase was also considered an early phase. In reexamining Meggers and Evans' seriation the developing decorative techniques and vessel forms of the two ceramic complexes—Acauan and Mangueiras—would fully flourish during the Marajoara phase.

## The Initial Period of Social Complexity

Although both the Acauan and the Mangueiras phases have not been securely dated to the period immediately before or contemporaneous with the Marajoara phase, the Formiga phase is clearly part of the initial period that characterizes the development of regional societies on Marajó Island.

Formiga villages consisted mainly of small mound groups located on the *campo*, close to both a water source and patches of forest. Seven sites have been identified, comprising one to seven mounds each. These mounds have areas that range from 79 to 3,356 square meters, with heights between 0.5 and 1.25 meters. Evidence of the artificial construction of an earthen platform prior to the occupational levels has been found at two sites (Coroca and Mucajá). In the latter, remnants of a well have also been identified, probably because the nearby stream would not retain enough water during the summer months—another common trait with the later Marajoara phase, when artificial ponds were widely used. Formiga phase ceramics, however, are mostly plain, the most popular decorative technique being brushing, although an incised type resembled the Ananatuba phase Sipó Incised type. Meggers and Evans (1957:241) considered that Formiga phase sites were "more isolated and perhaps also less receptive to ceramic innovations than those of the Mangueiras phase."

Contact with the Marajoara phase was established by Marajoara phase sherds (excised and incised types) found in two Formiga phase sites. Meggers and Evans considered the Marajoara phase sherds found on the surface of the Formiga site and on both the surface and the upper levels of the São Leão site as "trade sherds" (Meggers and Evans 1957:240; Simões 1967). Eroded Formiga phase ceramics were also found at Ilha do Fogo, a Marajoara mound (Schaan 2001b). Burned bones without any associated grave paraphernalia were found at the Formiga site. The beginning and duration of the Formiga phase was dated by the thermoluminescence method on pottery sherds excavated by Simões from mounds located on the southeast (AD 10–400), and on sherds from northern Marajó at the Formiga site (AD 610–837), collected by Meggers and Evans (Meggers and Danon 1988). Roosevelt (1991:64) was unconvinced by Meggers and Evans' seriation, and considered that in some cases Formiga and Mangueiras sherds might actually be domestic pottery from Marajoara phase seasonal occupations.

The fact is that investigations of these earlier phases consisted mainly of the collection of ceramic sherds from small test excavations, without in-depth study of larger site areas. These early occupations may have contributed more to the development of social complexity than researchers believed previously.

When considering the changes in settlement patterns through time, early occupations may have begun to develop knowledge of the natural environment, learning resource exploitation techniques that would be critical for the development of social complexity. The main evidence for this is the location of Formiga phase mounds that were intentionally built on the inundated savannas, suggesting that the members of these simple societies were already experimenting with intensive fishing before regional societies appeared.

Other indications of the relationships between the Marajoara phase and previous occupations are provided by the analysis of their ceramic industries. Vessel forms, ceramic technology, and decoration techniques that became important during the Marajoara phase as prestigious items owned by the elite display technology present in earlier phases.

Regardless of what the ethnic configuration might have been during the Formiga phase, the Marajoara culture was not a new group of people who arrived on the island, but the outcome of a successful local subsistence strategy, coupled with new forms of social organization and religious systems, which evolved and spread over the island within a few centuries. It was the result of local sociocultural changes, inaugurating a new way of life.

## Cultural Change and Landscape Modifications at the Mouth of the Amazon River

The archaeological record indicates that native Amazonians largely dealt with ecological conditions by taking the most advantage of them, changing landscapes in diverse ways to accommodate their needs. Therefore, instead of looking at the archaeological record as a patchwork of ceramic cultures, it is more enlightening to look at the ways those populations interacted with the environment. People choose places to live for particular reasons, and their strategies include concerns about the quality and availability of natural resources. As we look at how landscapes change through time, how the spatial organization of settlements change, and how social relations are formed in a given territory, it is possible to see these strategies imprinted in the landscape.

A comprehensive picture of Marajoara phase site distribution is biased by insufficient research. No complete archaeological survey of the Marajó savannas has been ever attempted. This is a huge rural area with no roads and few permanent rivers, divided up into a number of large ranches with low population density. Our present knowledge of site distribution is thus biased by a lack of systematic research, and based solely upon limited information gathered from the isolated reports of scientists, explorers, and looters. The actual number of Marajoara mounds is likely to be at least 10 times greater than currently assumed.

The only information available on settlement patterns on the eastern savannas comes from data from the Marajó Project, which investigated the upper courses of the Camará and Goiapi Rivers (Simões 1967). Both rivers are formed by numerous small streams, locally called *igarapés*. This 450-square-kilometer area is a flat savanna dotted by forest islands, most of which are archaeological mounds (Figure 2.4). Research at those sites has been limited, with the exception of Teso dos Bichos and Fortaleza, which were more thoroughly excavated.

Investigations undertaken by the Marajó Project consisted in locating and measuring site area and elevation, collecting sherds and artifacts from the surface, and excavating small test units. The sites belonged to the Marajoara and Formiga phases. Contact between the two phases was established based on a few Marajoara phase pottery sherds found at the upper levels of Formiga phase mounds. Ceramic data were used to

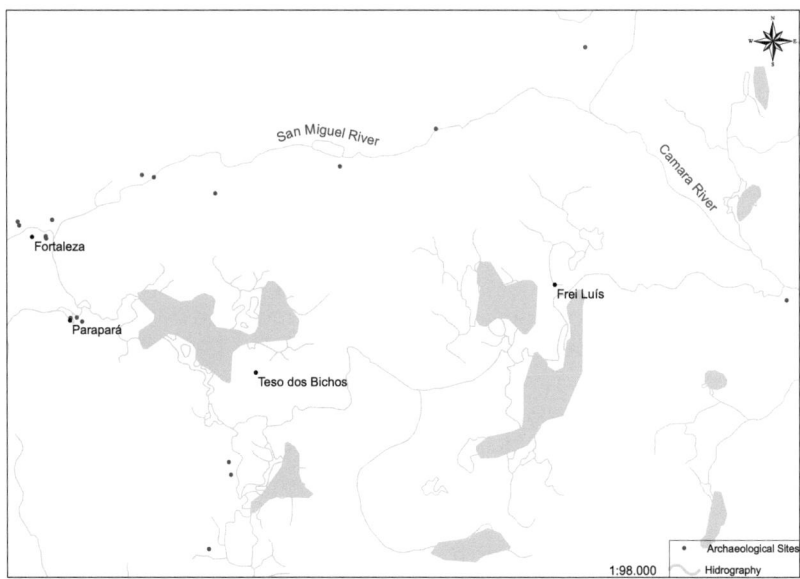

Figure 2.4.   Upper Goiapi River sites.

establish relative chronology following the Meggers and Evans seriation system (Simões 1967, 1969). The reports (Figueiredo 1963; Figueiredo and Simões 1962; Simões and Figueiredo 1965) are sparse as far as cultural features and stratigraphy are concerned. No effort was made to understand the distribution of mounds and the relationship between them from a regional perspective.

By examining data produced by this research, we can conclude the following: (1) ceremonial mounds are usually associated with two or more habitation mounds; (2) ceremonial mounds tend to be located along the banks where a stream meets the main river rather than along the main course of the river; (3) the Formiga phase habitation mound cluster is flanked by two Marajoara phase ceremonial mounds; and (4) some of the Marajoara phase mounds (such as Parapará, Salitre, and Teso dos Bichos) were built on the streambed, suggesting that they initially functioned as dams.

The largest group of mounds in the area is the Fortaleza site. Some of the 14 mounds of the Fortaleza site were excavated by Hartt (1871) and Farabee (1921). They reported that only one of the mounds was a burial site, whereas the others were used for habitation.

According to Farabee, funerary practices included secondary inhumation as well as cremation. Few items were included inside the burial urns. Tangas (female pubic covers) were occasionally found, but more often they were placed outside the urn, along with ceramic plates (Meggers and Evans 1957:306).

Absolute dates for the Marajoara phase occupation in the area come from the Frei Luis mound. Frei Luis is a small mound measuring 75 by 60 meters with a maximum height of 1 meter located at the Frei Luis stream, a tributary of the upper Camará River. A test excavation at the top of the mound indicated that the deposits were 60 centimeters deep. Below that level, they uncovered human bones, charcoal, abundant large sherds, and a small jar inside of an urn dating between AD 480 and AD 580 (Simões and Figueiredo 1965). A thermoluminescence date from ceramics found at the same site showed that the Marajoara phase occupation may have begun as early as AD 220 (Meggers and Danon 1988).

The data produced by the Marajó Project indicated population aggregation at the headwaters of both the Goiapi and Camará Rivers, with most of the mounds found at the former. The ceramic evidence indicates a Formiga phase occupation earlier in the area. When Marajoara phase pottery appears in the archaeological record at the Ilha do Fogo mound, it becomes increasingly popular, eventually replacing Formiga phase vessels (Schaan 2001b).

In 2007 and 2008, I conducted a survey of all mounds in the upper Goiapi River region, although I was unable to determine exactly which mounds comprised Farabee's 14-mound cluster, as he left no map. At present, only two mounds (now called Casa Velha 1 and 2) are located on the Fortaleza ranch. The site of Casa Velha 2, however, is a ceremonial mound surrounded by several habitation mounds.

Mound distribution at the upper Goiapi River conforms to the general rules for the Marajoara phase settlements: location at river headwaters, close to lakes and small streams, habitation mounds outnumbering ceremonial mounds, and ceremonial mounds near or on top of dams. The mounds of the Casa Velha site, located next to the Parapará Lake, are flanked by a huge pond. Here, we excavated an anthropomorphic burial urn (Figure 2.5) that was partially uncovered by erosion. This excavation, at the edge of mound, gave us the opportunity to observe that local stratigraphy is similar to that of the Camutins mounds and other ceremonial mounds described in the literature.

In the upper Goiapi River region, the most important ceremonial mound was probably Teso dos Bichos, studied by Roosevelt in the 1980s, where she found that the top 2 meters of deposits were already gone due to excavations and looting that have taken place since the nineteenth century.

Figure 2.5. Excavation of an anthropomorphic burial vessel at the Casa Velha 2 site (photographs by Denise Schaan; drawing by Deise Lobo).

The site was first excavated in 1871 by J. Steere, an Italian naturalist. He collected jars, pots, and many ceramic objects that were sent to the Museum of Anthropology at the University of Michigan (Meggers 1945). Steere reported funerary urns containing deteriorated bones and tangas, covered by upside-down flat bowls. He noted that the deposits comprised three different strata, at least two of which were separated by a layer of burned clay, charcoal, ashes, and broken pottery. He also stressed that the burial urns showed decorative differences between strata, suggesting that they belonged to three different occupations or cultural periods (Palmatary 1950:271). This pattern was confirmed by Ferreira Penna (1877), who excavated a collection at the same site for the National Museum of Rio de Janeiro.

A number of anonymous looters explored the site after Steere and Ferreira Penna. The constant trampling of cattle over the mound and seasonal rainfall caused the erosion of its flanks, and looted trenches led to the reduction of its original height. The Teso dos Bichos site was excavated again by archaeologists from the Goeldi Museum during the dry seasons of 1962 and 1964 (Corrêa et al. 1964; Figueiredo and Simões 1962); they reported 10 different strata, including layers of baked clay, sterile sand, and charcoal at the 4- to 5-meter-high mound. Artifacts included a snuffer, a pottery stool, and an axe.

In 1977, Alves and Lourenço (1981) from the Pará Federal University carried out a geophysical survey at Teso dos Bichos, applying the magnetic and electroresistivity methods, in a 140-meter by 120-meter area. Investigating the source of the anomalies, they found a set of six kilns (Roosevelt would later call them stoves) in the first excavation, distributed in different levels. In the second excavation, they found a funerary urn containing bones, located under a thick layer of burned clay, as well as a ceramic kiln similar to the previous ones. The excavations also produced tanga fragments, pottery stools, vessels, and a variety of pottery sherds.

Alves and Lourenço, however, point out that not all the anomalies with a similar pattern in the geophysical map were created by kilns: "We found out that many of the anomalies that we believed were produced by kilns were, in fact, cavities filled with humus rich material" (1981:8). They also emphasized the fact that the magnetometer may assign similar measurements to different types of anomalies: "The amplitude of magnetic anomalies depends on the depth and extension of the source. A narrow anomaly originates from a superficial source, while a large anomaly can be generated by a deep source as well as a source that, even when superficial, has a large horizontal extension" (Alves and Lourenço 1981:14).

In two seasons between 1983 and 1985, Roosevelt and her multidisciplinary team carried out an intensive geophysical survey that covered the 5-acre surface of the Teso dos Bichos site, investigating archaeological and geological features using four different geophysical methods. According to Roosevelt, the geophysical methods provided complementary information: the magnetometers were able to detect the stoves, whereas the resistivity survey provided information on stratigraphy (Roosevelt 1991:193).

Baked clay stoves, earthworks, and garbage deposits were the main archaeological features found. According to Roosevelt (1991:211), the most intense magnetic anomalies were located in an oval area at the center of the site. Differences in intensity between geophysical anomalies were reportedly due to the number and depth of stoves, rather than the type of feature. The stoves were reportedly grouped in sets of 6 to 12 units. Their east-west orientation was believed to indicate a similar orientation for the houses.

Roosevelt also identified different kinds of refuse related to domestic activities. She states that some primary deposits could be identified: "[They may] consist of a thin layer of ash and carbonized plant remains that were blown from a hearth by the wind and came to rest on a house floor. Another kind of primary deposit consisted of layers of spent fuel, bones, and offal possibly left over after a feast." "Caches of broken ceremonial pottery and tools buried under minor earthworks" were also reported as being a type of primary garbage (Roosevelt 1991: 237).

Although no entire house floor was excavated, Roosevelt identified a few house features, such as decomposed adobe and post molds or post holes. Instead of finding linear sequences of post holes or post molds, however, post holes were found as single units, probably as a central structure in the house. "Most of the post features excavated near hearths at Teso dos Bichos seem to have been roof or rack supports rather than walls" (Roosevelt 1991:239).

The most discernible features identified by Roosevelt are the stoves. The stoves are described as "U-shaped troughs, built of hand-plastered, ceramic quality clay into prepared pits or ditches, with the stove walls projecting about 10-20 cm or more above the ground" (Roosevelt 1991:254). All of the stoves reportedly belong to the Pacoval subphase (ca. AD 700–1100), which comprises the middle 2 meters of deposits in the mound, where the major magnetic anomalies were detected. These anomalies were groups of stoves lying between 0.5 and 1.5 meters below the surface (Roosevelt 1991:242–243).

Criticisms of Roosevelt's interpretation of the stoves (Barse 1993; Meggers 1992) are based on her assumption that unexcavated geophysical anomalies allow for the calculation of the number and size of the houses. Geophysics indicates anomalous areas but hardly tell archaeologists what is buried below the ground. Based on the number and orientation of the stoves, Roosevelt believes that the village was composed of

20 or more multifamily elongated houses. These communal houses were supposedly inhabited by 25 people on average, leading to a population of 1,000 people for the site. Houses would have been located around an open plaza (Roosevelt 1991:401).

Based on several radiocarbon dates, Roosevelt estimated that the Pacoval subphase lasted from AD 700 to AD 1100, while the Teso subphase was dated from AD 1100 to AD 1300. The main period of occupation dates from AD 800 to AD 1100.

## Research at the Camutins Site

To investigate a discrete settlement system and address questions related to the Marajoara social organization and landscape transformation, I conducted archaeological investigations at the upper Anajás River between 1998 and 2002, in an area located at the border between the savanna and the forest in the center of the island (Schaan 2004, 2008). The 34 mounds unevenly distributed along the *Igarapé dos Camutins*[1] (Figure 2.6)—a right-bank tributary of the upper Anajás River—is the largest mound group in the island.

The research consisted of surveying the entire Camutins River, locating and mapping the mounds and other landscape features such

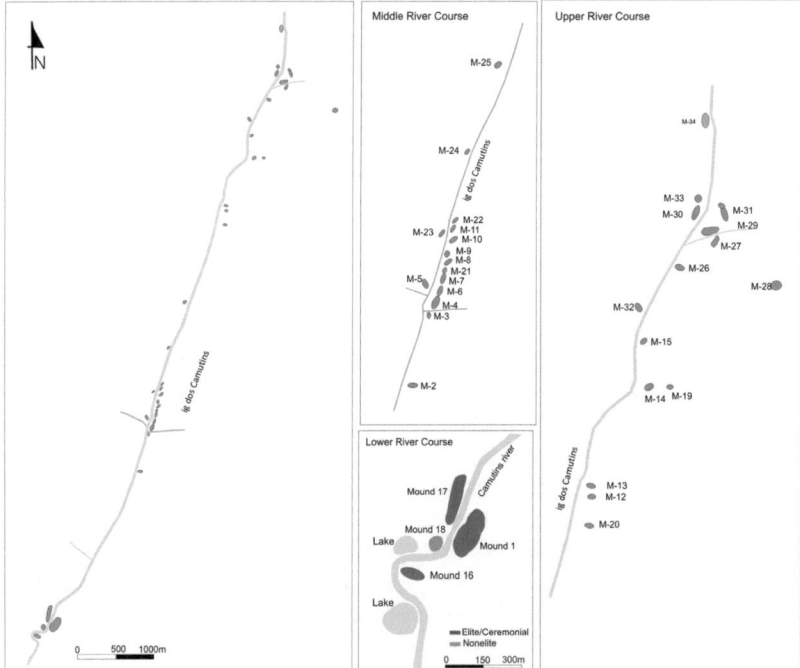

Figure 2.6. Distribution of mounds along the Camutins River.

as ponds and dams, making surface collections, and excavating the two major ceremonial mounds, M-1 and M-17. Around the mid-twentieth century, Meggers and Evans excavated M-1, M-17, and M-14, and Hilbert and Shultz excavated M-30 and M-34 (two upper Camutins River ceremonial mounds now leveled by looting).

Although its geographical circumscription suggests autonomy, the Camutins site cannot be fully understood within its own boundaries. For example, similarities in funerary practices and pottery styles between Camutins site mounds and other sites located along the Anajás River and its tributaries suggest that the area of influence of the Camutins polity extended further.

Between 1960 and 1980, Camutins site mounds were intensively looted by landowners, who invested considerable time and resources collecting archaeological ceramics to sell. Although the extent of the looting is unknown, at least 5,000 ceramic and lithic objects from the site have been incorporated into private collections in Brazil, the United States, and Europe. Most of the artifacts that have remained in Brazil are part of museum collections in Belém, Rio, and São Paulo. Unfortunately, decades of intense looting, trampling of cattle, and erosion have caused extensive damage to the mounds. With the exception of the Belém mound (M-17), which was preserved by the landowner, ceremonial mounds have lost between 40 percent and 90 percent of their archaeological deposits.

Besides site preservation issues, both the importance of the Camutins site on pre-Columbian Marajó Island (given its monumentality and range of influence) and the possible loss of the remaining archaeological patrimony in the coming decades were important in considering the feasibility of the research.

The Anajás River has its headwaters on the western border of the *campo*, meeting areas of forest as it flows southwest. Most of the right-bank tributaries of its upper course (including *Igarapé dos Camutins*) drain mixed areas of both forest and *campo* vegetation. The area drained by the Camutins River, for instance, is characterized by *campos* (especially at its upper course) as well as some patches of gallery forest. The mounds, located along the riverbanks, are covered with the typical vegetation found in Amazonian archaeological sites, such as palms and various fruit trees.

*Igarapé dos Camutins* is the first right-bank tributary of the Anajás River. It runs almost as a straight line from north to south, with more pronounced curves in its lower course, between the area where the higher elite/ceremonial mounds are located and its mouth at the Anajás River, 3 kilometers below. The permanent connection it provides between the Camutins and Anajás Rivers, even during the dry season, makes it the main route for contact between populations living along the Camutins River and other sites in the Anajás River basin.

At the peak of the rainy season (March to May), water levels typically rise 4 meters above normal. As a consequence, the Camutins River shoreline disappears, and no visible limits remain between the riverbed and the flooded *campo*, forming a huge lake that spreads as far as the eye can see. Above this shallow body of water, only forest islands (most of which are archaeological mounds) are visible.

Due to river overflow and the accumulation of rainwater, a multitude of small canals and water courses are formed, linking several large and small rivers that are not usually connected. From February to July, the upper course of the Camutins River, which remains isolated in the dry season, connects to other rivers through a narrow, shallow canal known as Igarapé do Urubu, located on its left bank. Igarapé do Urubu provides a transportation route from the Camutins River to the Anajás-Mirim Creek and eventually to the large Arari River, which flows into Marajó Bay. In the past, this route provided an important aquatic connection between Camutins and other polities located on the eastern *campos*.

The Camutins River, as is true with most rivers on the island, does not have a permanent water source; therefore, the volume of water dramatically and quickly decreases at the end of June, which marks the beginning of the dry season. Nowadays, the receding waters, especially in the upper course, require prompt management to guarantee the survival of the local population and livestock. The area is sparsely inhabited; a few local cowboys and their families run ranches for landowners who prefer to live in the city of Belém. Each year, a number of earth dams are rebuilt in several locations along the river. These small dams (*barragens*) are constructed at different times of the year, depending on the local needs as the season progresses. The work consists in filling the riverbed with sandy clay excavated from the riverbanks.

During the dry season, when parts of the riverbed are exposed, pre-Columbian water management features are observed that attest to a long history of water control. As soon as the dams are built, narrow, elongated lakes filled with shallow and stagnated water are formed. By the end of the season (November or December), malnourished water buffaloes often find themselves trapped in the muddy shores, eventually dying and contaminating the water. This situation is especially dramatic along the upper course, where poor management, trampling by buffaloes, and erosion obstruct the river and cause even drier conditions early in the season. According to Hilbert's informants in 1950, in the past the river had a deeper bed, flowing constantly even during the summer months. Fish were abundant, and included large species (Hilbert 1952:16–17), which have all but disappeared.

In the lower course, dry conditions arrive late in the summer, but problems with navigation begin as early as in August. Every year a *tapagem*[2] (wall dam) is rebuilt some 500 meters upstream from the river delta. From that point, it retains water 2.5 to 3 kilometers upriver, reaching the mounds located at the lower course. In this stretch of the river, an abundance of fish is available for at least three to four months (July to October). However, the situation changes from year to year, depending on management and the severity of the season. For example, by the end of a exceptionally dry summer in 2002, the riverbed in front of M-1 and M-17 was exposed. An emergency dam was then built just below M-16 to retain water for cattle.

Although in recent times the environment has suffered enormously due to buffalo and cattle ranching (causing river obstruction and the erosion of riverbanks and mounds), seasonal variations in water levels surely affected pre-Columbian populations as well. Indeed, current problems may very well contrast with ancient strategies of landscape management that were in place to guarantee water availability and abundant fish resources.

### Landscape Archaeology at the Camutins Site

The survey along *Igarapé dos Camutins* was conducted in three field stages: March 1999, July 2001, and September and November 2002. In 2002, the exposure of the *campo* and of parts of the riverbed allowed for the observation and mapping of features such as pond excavation, earthen bridges, and wells, which are not visible during the rainy season. The river itself was mapped by GPS, since the available maps are not accurate.

The surveys located 18 of the 20 mounds described by Meggers and Evans (1957:279–295), all the eight mounds reported by Hilbert (1952:10–15), and six other previously unreported habitation mounds. In addition, we located a number of other interesting elements, such as artificial excavations of ponds used for fish farming and water management adjacent to 10 of the mounds (M-2, M-5, M-7, M-9, M-10, M-13, M-16, M-18, M-24, and M-25), with areas that ranged from 18 to 13,576 square meters (the largest one represents the removal of at least 27,000 cubic meters of sediments); four sources of spring water or natural wells (close to M-1, M-14, M-18, and M-27); five canals or streams; and five modern dams (one *tapagem* and four *barragens*). Three main groups of mounds were identified along the lower, middle, and upper courses of the river (Figure 2.6.), indicating a hierarchical pattern contingent upon geographical, ecological, and cultural variables.

Mounds located along the upper course of *Igarapé dos Camutins* (especially M-30, M-31, and M-34) were found to be in poor condition from looting and erosion. M-14 was flattened to build a dam and a landing strip in 2000, and M-19 was flooded by the resulting lake. Data gathered from research conducted in 1949 and 1950 (Hilbert 1952; also reported by Meggers and Evans 1957) was incorporated to expand the database and better understand the settlement structure.

During the surveys, we collected sherds from the surface of the mounds. Following Simões and Figueiredo (1965), we considered, based on artifact analysis, the existence of two functionally distinct types of mounds: habitation and ceremonial/elite housing mounds. Habitation mounds typically have a low proportion of decorated pottery (0–5.9 percent). Ceremonial mounds tend to have a higher proportions (around 12 percent on average) and greater diversity of decorative types.

The mounds along the Camutins River were found in three main clusters. The first cluster consisted of four mounds located along the lower course of the river. M-1, M-16, and M-17 contained burials and decorated pottery, whereas M-18 contained only plain sherds. Mound area and height suggest a rank/size hierarchy: M-1 would be first in importance, M-17 would be second, and M-16 and M-18, by virtue of being smaller and situated next to the other two, would be third and fourth in the hierarchy.

The second cluster consisted of 15 mounds located along the middle course of the river, with heights varying from 1.5 to 5.44 meters, and areas ranging from 17 to 959 square meters. Four of the mounds were on the right bank, but most were on the left (eastern) bank. Small artificial ponds were identified next to mounds M-2, M-7, M-9, M-10, M-24, and M-25. Surface collection produced a majority of plain sherds from coarse-finished, thick-walled vessels. Many mounds have an uneven surface, indicating multiple platforms at different levels. Some were actually two separate mounds linked by an earthen causeway that was frequently lower than the mounds themselves.

The 15 mounds comprising the third cluster were located along the upper course of the *Igarapé dos Camutins* and were more intensively affected by looting, erosion, and trampling of water buffaloes. These mounds were not as high as the ones on the middle course, which may suggest that they were occupied later in the sequence. Hilbert (1952) observed that most of the mounds he visited remained underwater during the winter, which he credited to ecological changes caused by buffalo ranching.

I identified three elite and 12 habitation mounds along the upper course of the river. The elite mounds have areas that range from 600 to 1,360 square meters. They may originally have been 2 to 3 meters high, but are now flattened by looting and erosion. The habitation mounds are 17 to 2,576 square meters in area, and their heights vary between 1 and 7 meters. The habitation mounds differ from the elite mounds in the type and proportions of artifact remains as well as cultural features. Mounds with burials and a high proportion of decorated sherds were considered elite mounds, whereas mounds without burials and few or no decorated sherds were interpreted as habitation mounds for the nonelite population.

A single artificial pond was identified next to M-13. M-26 is located on the riverbed, and was possibly originally constructed as a dam. Springs were identified next to M-27 and M-29.

The dynamic nature of the island's hydrography as well as constant rebuilding of dams by ranchers to retain water for livestock makes it difficult to determine that dams without potsherds date to pre-Columbian times. I believe that modern populations are replacing native strategies for landscape management, as Smith (2002) has also observed. Small earthworks not related to any specific mound, such as seasonal *barragens,* were built in the last century.

Borrow pits or ponds found next to a number of mounds are clearly related to the removal of sediments both for mound construction and water management. These ponds are connected to the river during part of the year. Wooden and earthen dams might have been used in conjunction with the ponds, channeling fish into permanent and deeper bodies of water, which were separated from the river at low tide. There, aquatic fauna would be trapped and could be easily harvested.

Two large ponds (5,300 and 13,400 square meters, respectively) were found next to mounds along the lower course of the river (Figure 2.6). These ponds have been partially filled with sediment deposited by the river during the rainy season, but they are still some 2 meters deep in relation to the surrounding terrain. Below the fine silt that fills the ponds, there is a 50- to 70-centimeter-thick clay substratum, indicating that initially the ponds were at least 2 to 3 meters deep.

M-16 may have originally been a dam, used in conjunction with the ponds for water management and fish farming. The movement of aquatic fauna between the river and the ponds could have been managed through removable fences, thus securing live protein for consumption throughout the summer months. At the peak of the dry season, the ponds would be completely isolated from the river. This system would require annual maintenance to remove sediments deposited in the ponds by the floodwaters. The removal of sediments to nearby areas contributed to mound construction. The initial excavation of the two ponds likely required a communal effort over a short period of time, perhaps a year or more. Routine maintenance, on the other hand, would be seasonal, requiring less labor investment. If the ponds were small in the beginning of the occupation, their present size may represent incremental work over several decades or centuries.

Excavation rates can be measured by a study of mound construction. Thick layers of silt added over extensive areas of the mounds are related to more intensive excavations of the ponds. Conversely, discrete episodes of mound building, with thin layers of silt between occupation strata, correlate to seasonal maintenance of the fish farming system. Episodes of mound construction are thus a testimony of earthmoving activities, and can provide a chronology for water management and the mobilization of labor.

Similar construction might have taken place in the upper course of the river, but the dramatic impact inflicted by modern land management on the area has rendered it difficult to discern landscape features there.

In most mounds, an initial layer of sediment was laid down to build a higher platform intended for habitation (Meggers and Evans 1957:399). Some additional layers were naturally deposited during the occupation, and others were purposely added either to dispose of sediments removed from ponds or to intentionally raise the elevation of these platforms. Although the intensity of mound building varied through time, episodes of mound construction were probably seasonal, and consisted of adding layers to limited areas of the mound, rather then its entire surface. This can be confirmed by the existence of platforms at different heights in some of the individual mounds, as well as by an analysis of M-1 and M-17 mound-building episodes, which were supported by radiocarbon dates. Some elongated mounds were initially two mounds linked by an earthen causeway. As the population grew, new mounds were built attached to existing ones; this second mound is typically smaller and lower than the initial one.

The morphology of the mounds only allows for linear villages, with houses placed along the riverbanks, as is common among riverine populations dependent upon fishing. Linear villages were reported among the contemporary *Karajás*, and the ethnohistorical *Omáguas* (Costa and Malhano 1986:30). In these linear villages, houses are usually rectangular or oval. Most of the mounds have a top surface no wider than 20 meters, which would suggest the existence of single long houses with dimensions no greater than 15 to 20 meters wide and 25 to 30 meters long. Based on the excavation of stove groups at the Teso dos Bichos site, as well as on a study of magnetic anomalies that were interpreted as stoves, though not all of them were excavated, Roosevelt (1991:336) suggests the existence of long communal houses similar to the ethnographic *maloca*.

Although definition of the "shape, size and orientation of the dwellings" requires further excavations, Roosevelt (1991:335) suggests that Marajoara phase houses measured 30 meters by 20 meters. No complete house floor has been excavated. Moreover, besides the fact that all researchers have focused their attention on the study of elite/ceremonial mounds, there is very limited knowledge about the structure of habitation mounds. For these reasons, house size estimates are based upon the available flat surface on habitation mounds as well as on the assumption that they were all used for habitation purposes. Therefore if all the Camutins River mounds were contemporaneous, these figures would indicate a minimum of 28 houses and a maximum of 43 houses at any given moment.

Based on the available data, population figures are difficult to assess. Roosevelt (Bevan and Roosevelt 2003:329; Roosevelt 1991:342) has estimated a population of 1,000 people for the Teso dos Bichos mound, and 78 to 156 people for the Guajará mound.

Ethnographic analogy suggests that communal *malocas* measuring 20 meters by 30 meters were inhabited by 3 to 12 nuclear families (Costa and Malhano 1986; Jackson 1994). Roosevelt (1991:342) estimates that these dwellings would house between 35 and 60 people, with an average of 40 people. With a conservative perspective, it seems reasonable to assume that population figures for each house ranged from 12 to 60 people. Therefore, based on mound area and on ethnographic information regarding the average number of inhabitants per house, the 28 Camutins habitation mounds would have supported a total population of 1,660 persons. For the elite mounds, I have estimated a population of 380 people. Therefore, the community would not have had more than 2,000 inhabitants at any point in time.

Data collected during the surveys allowed for the assessment of the distinctive settlement pattern displayed by the Camutins site. The spatial location of the mounds is interpreted as reflecting social distance, differential access to natural resources and exchange routes, and defense strategies.

A spatial analysis of the Camutins site indicates that habitation mounds are arranged in a linear pattern, within an area flanked by elite mounds. In discussing rules for linear settlement systems along rivers, Flannery (1976:180) proposed that villages are likely founded both upstream and downstream, symmetrically away from a center, whereas new daughter communities tend to be created in the spaces between them. For the Camutins River mounds, variables such as resource proximity, social distance, administrative functions, and political control need to be examined to account for settlement location and the relation between mounds.

The two largest mounds (M-1 and M-17) are located at the lower course of the river. The largest mound in the settlement (M-1) has an area of three standard deviations above the mean. The second largest (M-17) has an area between one and two standard deviations above the mean. The third largest mound size has an area one standard deviation above the mean, and is represented by six mounds (M-16, M-18, M-29, M-30, M-31, and M-34). The remaining mounds fall into the fourth category, with areas below the mean (Figure 2.7). The first category is significantly larger than the second. The third category includes not only elite mounds, but also two habitation mounds that are located close to elite mounds. Thus, habitation mounds are only large when they are associated with elite mounds.

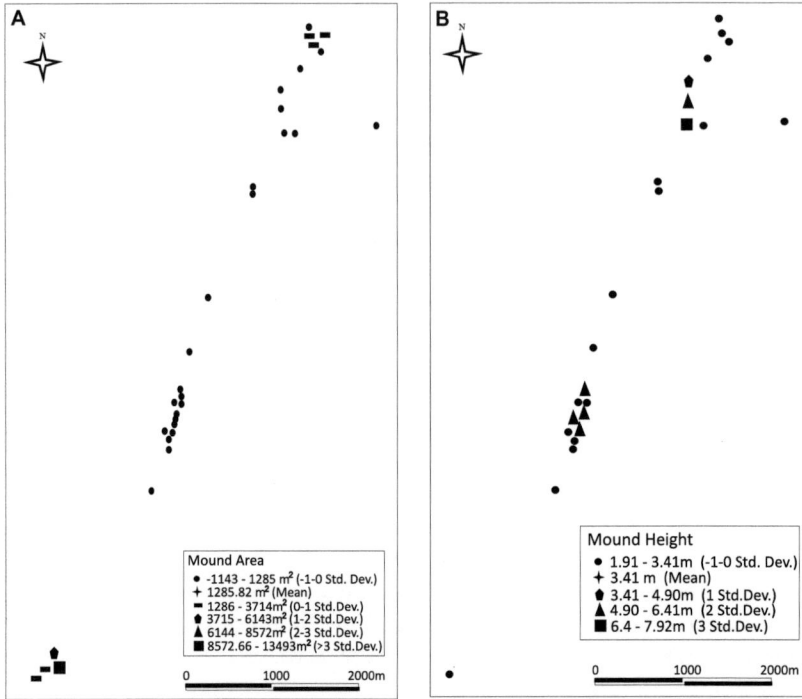

Figure 2.7. Distribution of mounds according to area (A) and height (B). After Schaan (2004:176, 179).

As a result, I defined four major settlement categories:

- Type 1: M-1, a political and elite/ceremonial center, on the lower course of the river;
- Type 2: M-17, an elite/ceremonial mound, on the lower course of the river;
- Type 3: elite/ceremonial mounds or habitation mounds associated with an elite mound, being two mounds along the lower river course and four mounds on the upper river course; and
- Type 4: habitation mounds located along the middle and upper courses of the river.

The rank-size rule is applied to population sizes; city sizes around the world show consistency, and for this reason the rule is applied to archeological phenomena. Cities that concentrate higher numbers of administrative offices and economic functions tend to have higher populations and sizes. Based on rank-size, we can assume economic and/or sociopolitical relevance.

According to the rank-size rule and function, the Camutins settlement is hierarchical, exhibiting the following characteristics: (1) a political, ceremonial, and administrative nucleus; (2) secondary and tertiary villages with ceremonial structures; and (3) a village group without ceremonial facilities. Applying Flannery's prediction for the evolution of linear settlement systems, secondary villages were away from the center in the first stage of expansion, while in a second stage, the gap between them would be filled according to social spacing rules. According to the Camutins data, the habitation mound village is distributed evenly between the elite/ceremonial group and the six elite "daughter" mounds.

The trend for the mounds in the middle course of the river located on the east side can most likely be explained by the proximity to the open *campo*. Based on settlement system rules for formative Mesoamerica, Flannery suggests that "local factors being equal, the side of the river chosen by a town may depend on where its more distant sustaining hinterland is located" (1976:174). We are not sure what kind of resources the east side offered. Although nowadays the area is not suitable for agriculture, cultivation remains a possibility with the appropriate management. Manioc agriculture is favored in partially forested areas so the plants are not totally exposed to the equatorial heat and sun. Another possibility is that this side of the river was better suited for foraging and hunting.

Another probable reason for the eastside location of the mounds is that by being located on the same riverbank, all of the mounds would be accessible on foot (Johnson 1972, 1977) because the construction of dams could have severed aquatic communication between groups of mounds over periods of time. Another important reason, highlighted by Farabee (1921), is that springwater sources are concentrated on the east side of the river.

Examining mound distribution according to height, a distinctive pattern appears (Figure 2.7). Elite mounds along the upper course are not particularly high; middle-course habitation mounds are higher than elite mounds in the upper course of the river. There is no correlation between height and function. For this reason, examining mound height within a same mound category, such as habitation mounds, should be instructive.

Examining habitation mound height, it is possible to see that higher mounds tend to be situated in the middle of the cluster, suggesting that clusters were growing internally from the center out. This would imply an internal dynamic, particularly within the middle course of the river habitation mound cluster, expanding both up and down the river through new mound construction, while at the same time increasing mound height from the center out. In sum, this settlement system displays a set of rules that may explain settlement evolution.

Accordingly, the Camutins River community could have evolved from a previous colonization of the lower course of the river, when some communal work was involved in the excavation of a pond and the construction of a dam—where M-16 is located—initiating the fish farming system. Since that area was under the control of the kin group founders, commoners who worked on the construction would have settled up the river within a distance that would guarantee proximity while, at the same time, maintaining an accepted social distance. As population increased, new hamlets would be constructed away from the ceremonial center downriver, thus expanding the settlement upstream. As the elite population grew, new elite residences were built, some close to the ceremonial core (thus the construction and occupation of M-17 and M-16) and some far away from the ceremonial center, in the community's upper river limits. The distance in this case may relate both to the genealogical distance (Service 1962) and to the need for an administrative control of the upper course of the river, where new hamlets were expanding. The upper course of the river was critical for water control, because if people there built a dam, they could jeopardize water availability in the lower course of the river. Placing an arm of the elite on the upper course of the river, therefore, was a matter of geopolitics, with a clear administrative and political function.

Ceremonies and feasting would be concentrated on the lower course, where fisheries were built. At the upper course of the river, local elite members were buried, and rituals were performed, thus justifying their claim to the territory through religious means. In controlling strategic areas of the landscape, the elite secured their access to resources.

### Research at M-1, the Ceremonial Center

The fieldwork at M-1 involved excavations of selected areas and cleaning of profiles in looted trenches (Figure 2.8). The mound was found to be badly looted and eroded, having lost about 50 percent of its archaeological deposits. The west slope of the mound, facing the river, was the only intact area. Trees are older there compared with the secondary vegetation that covers the rest of the mound.

Excavation of units and cleaning of profiles in different parts of the mound allowed for an understanding of its stratigraphy and rate of construction, since four radiocarbon dates were obtained from different

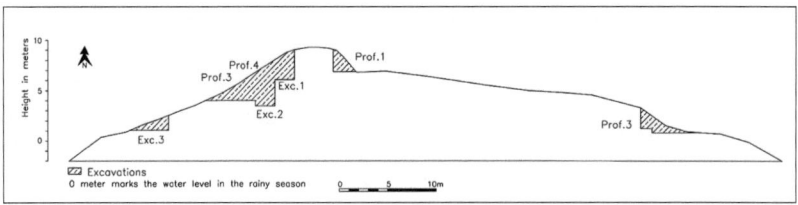

**Figure 2.8.** East-west cross-section of M-1. After Schaan (2004:186).

levels. Three profiles and two excavations showed a sequence of thick layers of light brownish-gray silt, followed by thin layers of charcoal, orange baked clay, and light greenish-gray sand. The thick layers represent massive episodes of mound construction through the addition of silt brought from nearby areas. The thin layers indicate discrete episodes of occupation. Hard clay surfaces may have resulted either from intentional firing to harden the surface, or as a result of hearths built on top of clayed sediment. Thin black layers are full of charcoal, probably residues of firing and sweeping.

The dates obtained at Profile 1 from a layer 115 centimeters below the surface (97.93-meter level), and from a charcoal layer located 210 to 213 centimeters below the surface (96.9-meter level), are dated to calibrated (cal) AD 785 to AD 995 and cal AD 645 to AD 880, respectively. These dates imply that in the highest area of the mound, about a meter of sediments was added over a period of roughly 130 years. A sample collected from the 93.83-meter level in Profile 4 nearby was dated to cal AD 660 to AD 790. The proximity of the two profiles and their stratigraphic similarity suggest that both were part of the same mound construction episode. Based on this understanding and on the fact that the 93.83-meter level in Profile 4 is technically contemporary to the 96.98-meter level in Profile 1 in radiocarbon years (with a 10-year difference according to the conventional radiocarbon age), I conclude that the 3.15-meter difference between the levels represents sediment that was added to the mound in a very short period of time, consisting of some 3,600 cubic meters of sediment added over an area at least 20 meters wide and 60 meters long. This implies that major mound construction efforts took place around AD 700.

A 2-meter by 2-meter excavation unit on the western slope of the mound revealed its most interesting feature: a 25-centimeter stratum of baked clay, 40 centimeters wide and 2.5 meters long, extending along the south-north axis, at the 95.14-meter level (Figures 2.9. and 2.10.A). This hard, flat, orange clay surface was created with clay and thick sherds, some of them with a partially vitrified external surface, indicating the use of high temperatures. Curbs 8 to 10 centimeters wide were elevated 2 centimeters along the structure. The orange clay surface was parallel to the excavated wall as well as the river. A fire pit containing sherds, burned clay, and charcoal was embedded in this sidewalk-like structure. The function of such a pit is not fully understood, but ethnographic analogy suggests the firing of small pottery vessels (see Rye 1981:99).

In the profiles on the west and east sides of the mound I found retaining walls consisting of a 10- to 20-centimeter-thick concave layer of large ceramic sherds and burned clay, which stretched from a northern lower elevation to a southern higher elevation, capping the light, sandy clay soil that constituted the mound (Figure 2.10B). These walls likely provided protection of the mound from erosion.

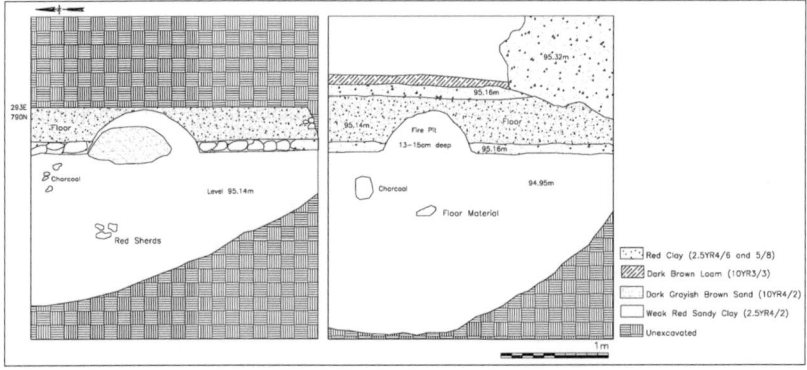

Figure 2.9. Top views of clay floor in M-1 Excavation 2. After Schaan (2004:188).

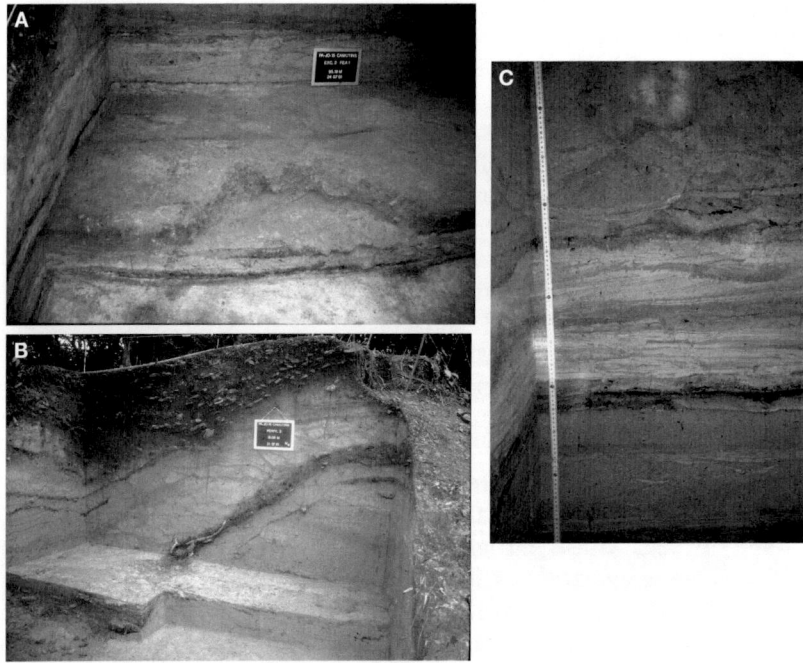

Figure 2.10. Features at M-1: (A) Excavation 2; (B) Profile 3; (C) Profile 4. Photographs by Denise Schaan.

Artifacts found during excavations at M-1 included an urn burial, fragments of decorated plates and vessels, fragments of tangas, stools, figurines, a fragmented stone axe (used as a burnishing tool), fragments of plain vessels related to food processing, remains of pottery production, and fragments of large ceramic plaques with leaf impressions, which probably served as a basis for pottery manufacture.

Based on the investigation, the mound was built through the addition of substantial layers of silt (Figure 2.10C) brought from areas located 50 to 300 meters away. Two such areas were identified across the river, interpreted as part of a hydraulic system aimed at intensifying fish harvesting. Periods of more intense excavation for the aquaculture system as well as seasonal maintenance are thought to have provided enough silt for building the four mounds that make up the ceremonial center. Between layers of sediment, thin layers of clay were laid to even and harden the surface, which, when burned, created superimposed orange and black layers. These thin layers represent very short episodes, since no other significant cultural features are related to them.

More extensive construction took place at the northeastern portion of the mound, as suggested by both the excavations and radiocarbon dates. Uneven layers of broken sherds and clay were added on the edges of this area as retaining walls to assure the stability of the structure. As the mound continued to be built, some platforms were eventually added, both to provide stability and create an even, hard surface. As argued here, radiocarbon dates and stratigraphy indicate that the mound was built during different episodes, which involved coordination.

Earthmoving activities per se do not imply mobilization of large labor forces, unless the work was done in a short period of time, thus involving a large number of people. Based on our excavations of the M-1 mound, four major building stages are identified, with different rates of construction.

Considering that the terrain may have had some elevation before construction was initiated, the mound reached a height of 5.83 meters around AD 700, suggesting that between AD 400 and AD 700, an area 50 meters wide and 80 meters long was occupied. This height was reached in an area 20 to 25 meters wide by 35 to 60 meters long. By AD 700, the highest platform was elevated 3 meters in a short time, which indicates the transportation and displacement of nearly 3,600 cubic meters of earth, which was probably carried in baskets. Sources for fill were located nearby, no farther than 300 meters away, where the two ponds are located. Transportation was accomplished using canoes, since the earth sources were located across the river.

In calculating the amount of people necessary for such an effort, a number of variables have to be taken into consideration, such as the number of working hours in a day and the number and size of loads that each person is able to carry over that period of time. Assuming that a single person could transport one cubic meter of fill per day (or 23 baskets of silt), it would have taken 3,600 days per worker to finish the construction. Considering that pond excavation and mound construction were seasonal activities (three to four months a year), that part of the mound could have been built by 40 people in just one season. Since the stratigraphy indicates the existence of short intervals (perhaps years) in between layers of added silt, and considering that probably other mounds were being built at the same time, it is unlikely that a workforce of more than 50 people was really necessary at any point.

After a period of more intensive mobilization of the workforce, construction at slower rates occurred in several parts of the mound. The difference between the two dates obtained from Profile 1 is 130 years, which is related to the addition of just a meter. Between AD 700 and AD 850, mound elevation increased at the riverside, whereas platforms were added at other levels and locations. A maximum height of 11 to 11.5 meters would have been attained only at the time of abandonment, ca. AD 1150 to AD 1320. Dates for the beginning of the occupation of the mound (AD 400), as well as its abandonment (AD 1320), are based on dates for the Marajoara phase (Meggers and Danon 1988:248).

Assuming that mound construction was linked to the creation of hydraulic works, it is reasonable to believe that by AD 700, a fishery that included the excavation of at least one fish pond and the construction of a dam was already in place. After that, less intensive mound-building activities indicate seasonal maintenance of the fishing facilities. It is unlikely that the initial excavation of the ponds and final implementation of an operating and successful fish farming system engaged a huge labor force, unless the work was accomplished in a very short period of time. In any case, these activities probably involved centralized coordination.

### Research at M-17, the Secondary Center

M-17 was investigated between September and November 2003. The objective was to investigate site function and determine its chronological position in the settlement system.

Excavation units initially measured 2.4 to 4 square meters, and were later expanded as needed. During the first stage, five excavations provided a good sample of household, cemetery, workshop, and open areas. During the second field stage, efforts were concentrated in the cemetery area, reaching an extension of about 25 square meters, where 24 burials were found. The excavation of burials and funerary patterns are discussed in Chapter 3.

Excavations at M-17 revealed a house floor area with an adjacent domestic garbage fill on the eastern slope of the mound (behind the house), a burial section, an area for pottery production, and peripheral areas (Figure 2.11). An excavation near the center of the mound top at an elevation of 8 meters revealed no prepared floors or hearths; in considering the abundance of pottery production remains and female objects (fragments of tangas; see Chapter 3), I suggest that this was an open area where pottery-making took place.

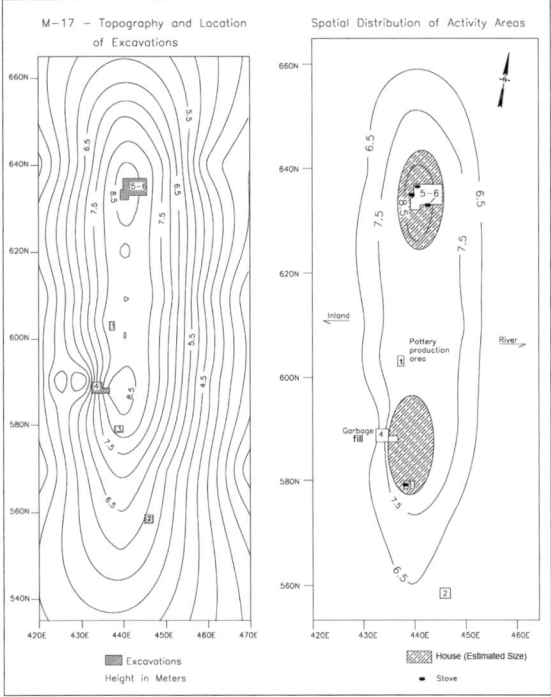

Figure 2.11. M-17 plan topographic map and location of excavations. After Schaan (2004:204).

Another excavation (1.2 meters by 2 meters) on the south side of the summit, at an elevation of 8 meters, uncovered the remains of two 40-centimeter-deep parallel ridges corresponding to the remains of a stove, although it differs from the other stoves found in the burial area, which are smaller (Figure 2.12B). Each of the two parallel baked clay walls (some 25 centimeters apart) was 10 centimeters wide and about 1 meter long. This structure is similar to the hearths found at the Guajará mound at the Monte Carmelo site (Bevan and Roosevelt 2003:318; Figure 15).

Below the stove remains, a charcoal layer was dated to cal AD 680 to AD 980. The next 2 meters of excavation uncovered a complex stratigraphy consisting of thick layers of light sterile sand alternating with thin layers of occupational floors. The composition of these layers was similar to those identified at M-1, most of which were compact with few sherds. As observed in M-1, the thin occupational layers were uneven and discontinuous, typically consisting of light brownish-gray sand, charcoal, and burned clay.

The complexity and unevenness of the occupational layers is likely due to the modification of the light sandy soil composition by biological

Figure 2.12. M-17 features: (A) Excavation 3, west wall profile; (B) Excavation 3, remains of a stove on the south wall and excavation floor; (C) "U"-shaped stove 1, Excavation 5; (D) Stove 2 in Excavation 5 wall.

debris (originally light gray sand acquires a darker color) and the use of fire (producing orange, hard clay, as well as charcoal layers). The thick light-sand layers were composed of light gray and light brownish-gray sediment, which resulted from the addition of silt from different parts of the river, thus accounting for the color differences. Two other charcoal layers were dated at the 160- to 170-centimeter depth (cal AD 660–900), and at the 250-centimeter depth (cal AD 720–740 and cal AD 760–960).

Fibrous charred wood found on a few charcoal layers was identified by Mário Jardim (Department of Botany, Museu Paraense Emílio Goeldi) as palm tree bark. During excavations, local workers identified them as açaí (*Euterpe oleracea*), and burned seeds were also identified by Mário Jardim as açaí, indicating its consumption by AD 700. This fruit is rich in protein and is important in the diet of contemporary Amazonian populations. The amino acid content of açaí juice is similar to that of an egg (Rogez 2000:175) (Figure 2.12A). A garbage dump was excavated on the central west slope of the mound, about 2 meters away from the flat summit. The top layers of this excavation consisted of ADE, which contained numerous large sherds as well as fragmented vessels. This type of soil results from the decomposition of biological debris (Kern and Kampf 1989; Smith 1980), indicating the presence of an organic garbage fill.

Due to the nature of the deposits and the artifacts they contained, this was considered a toss area where not only broken vessels but also organic matter was discarded. Underneath the ADE stratum, layers of yellowish-brown, sandy clay interspersed with layers of dark brown loam and charcoal indicated mound construction. Artifacts found in the garbage fill included numerous sherds in addition to two broken vessels containing wet, dark soil, bone fragments, and teeth. The vessels had been tossed from the mound top, and their contents had spilled or fallen out. One of the vessels contained a secondary inhumation, a funerary practice observed among ethnographic peoples in which human remains are kept in a vessel inside of the residence, instead of buried under the floor or kept in a mortuary temple (Morales-Chocano 2000:86–87; Schiffer 1987:83–84).

Artifacts found in this excavation included a fragmented seated figurine, two small undecorated bowls, abundant fragments of red and painted tangas, and a fragmented pot stand. Charred wood collected from a layer 85 centimeters below the surface was identified as açaí palm.

The excavations showed that M-17 was built through the addition of pale brown, sandy clay in a process similar to that described for M-1. However, M-1 had a number of platforms at different elevations, but M-17 had only one. During the last period of occupation, the activities were concentrated on a flat top area about 20 to 25 meters wide and 70 meters long.

The lower half of the mound consisted mostly of pale brown sand layers with few artifacts. An initial period of intensive mound construction was followed by seasonal construction. According to the radiocarbon dates obtained for three charcoal layers in the third excavation, 2 meters of sediments could have been deposited over a period of about 100 years, since dates range from AD 730 to AD 830. The AD 960 date obtained for Burial 18 may well represent the last occupation of the site. In this sense, the top 50-centimeter layer in Excavation 3 could have been added over another 100-year period.

Based on cultural features investigated (burials, garbage fill, stoves, and prepared floors), it is assumed that two houses existed on the mound. Stoves in the burial area (Figure 2.12C–D) indicate either a house where daily meals were prepared or a mortuary temple in which the stoves were used for ceremonial purposes, as ethnographic analogy suggests (Métraux 1947:26). Prepared floors in Excavation 3 and a nearby refuse disposal area identified in Excavation 4, consistent with a toss area at the back of the house, indicates a residence.

The chronology for the occupation of M-17 is a result of three radiocarbon dates obtained from charcoal layers in Excavation 3, and one radiocarbon date obtained from charred wood collected from the interior of Burial 18. The dates indicate that the top 3 meters of the mound were built in 300 years. Layers of thick silt strata with few artifacts in Excavation 3 indicate that a 1.8-meter layer (between the 5.96-meter and the 7.76-meter levels) was added during 100 years, indicating that every few years another layer of silt was added to the top of the mound.

The AD 900 to AD 1030 date refers to one of the last burials, marking the time of abandonment of the site.

Investigations at M-17 revealed a communal house, a mortuary temple, and a pottery production area, given the large amount of workshop waste found. A considerable amount of decorated pottery found inside and outside the mortuary temple was evidence of feasting and ceremonies.

The period investigated by this research spans from AD 700 to AD 1100, which covers the major development phase of the fish-farming system.

## A Model for the Development of Regional Polities on the Island

Data gathered during four years of investigation at the Camutins site and sites along the Anajás River made it possible to envision a long-term process of cultural change, encompassing the rise of complex societies and their demise before the European conquest. Based on archaeological evidence, I propose a specific sequence of events.

By AD 10, settlements associated with the simple ceramic industry of the Formiga phase are found in the seasonally inundated savannas. Formiga phase settlement patterns already displayed some of the features that would largely characterize the the Marajoara phase, such as an ellipsoid mound shape, mound building, and locations close to small streams in areas subjected to seasonal flooding (Myers 1973). The drastic landscape transformation conducted during the Marajoara phase was initiated by Formiga phase settlers. The development of a new subsistence system requires knowledge of the environment, and time for learning new strategies.

Mound construction was not a strategy to escape from seasonal floods, which affect only a portion of the island. People could have lived in noninundated areas, practicing horticulture, fishing, and hunting as during the Ananatuba and the Mangueiras phase occupations. The preference for flooded areas on the *campo* during the Formiga and Marajoara phases requires another explanation.

Formiga phase populations were probably exploring abundant fish resources at river headwaters, testing strategies that would fully develop during the Marajoara phase. Formiga and Marajoara ceramic industries apparently maintained independence, even when coexisting in the same sites. Similarities in ceramic technology among all the phases (e.g., crushed sherd tempering, same basic vessel shapes, some decoration techniques) indicate that they influenced one another. I call this period between AD 10 and AD 400 the "Incipient Marajoara."

By AD 400, dispersed communities aggregated into regionally organized societies: earthen mounds multiplied next to lakes and river headwaters, and a new ceramic style made its appearance.

Unlike their predecessors, the Marajoara phase populations did not look for permanent water courses, but chose to live in areas where the effects of the annual droughts would be more dramatic. Many Marajoara phase mounds are located in places considered uninhabitable today. Marajoara phase societies, however, managed to develop a complex culture away from large rivers and with little or no water from August to December. How would that be possible?

Understanding why complex societies emerged and developed on Marajó Island requires a correct assessment of its ecology and climate, and especially an understanding of how these factors affect both population dynamics and subsistence systems. Traditional ecological analysis of the island's landscapes frequently relied on the mistaken notion that the savannas (*campos*) were inundated *várzeas*, thus assuming that intensive crop cultivation was possible. However, the soil of the Marajoara savannas is clay-rich, nutrient-poor, and seasonally exposed to either torrential rains or unbearable equatorial heat (Organização dos Estados Americanos [OEA] 1974).

Therefore, the island's ecology constrains the development of large, complex, sedentary societies, unless such societies are capable of managing ecology to their advantage.

Native populations were aware of the dramatic seasonal fluctuations of resources. As contemporary populations still do, they built dams and ponds for managing the massive amounts of fish that reproduce in the headwaters and over the *campo*, where they fed off the floating meadows and the flooded forest every year. The fine silt removed during excavation of the ponds was used for building dams, and elevated villages, elite/ceremonial mounds, and causeways to connect them.

Landscape construction and fisheries have been identified at the Llanos de Mojos site in Bolivia (Erickson 2000, 2001), where the massive harvesting of aquatic fauna was possible through a complex system of fish weirs and fish ponds. The use of such subsistence strategies by indigenous societies would not be possible without an intimate knowledge of the environment.

On Marajó Island, indigenous populations may have seasonally moved to optimal areas to harvest, dry, and store fish. At first, families or communities may have cooperated by building dams when the waters started to recede. As the population grew, competition over these areas arose, and claims over the right to exploit resources may have generated conflict and division within groups. In this context, community leaders would have experienced a growth in power, with a strong economic basis to support sociopolitical institutions that involved permanent power and control.

At some point, fisheries came under the control of particular groups, who established the rules for access to resources. With time, the ability to control those resources in a symbolic way became particularly critical in determining the constitution and stability of the Marajoara polities. Even when the Marajoara elite did not manage to control fisheries by force, it could symbolically "control" fish reproduction by mediating the access to the supernatural world. The strong connection between resources and the spirit world is confirmed by the abundant ritual paraphernalia found in mounds located next to the major fish ponds.

According to Robert Carneiro (1970, 1981, 1998), areas with bountiful and concentrated resources would attract greater human populations, generating population pressure through time. He calls this "environmental circumscription," a key to understanding the timing of chiefdom formation. Population pressure in these ecologically circumscribed areas would lead to competition and warfare, and the stronger group would absorb the weaker. Carneiro hypothesized that the defeated group would either have to pay tribute or be enslaved by the victorious one. Later, Carneiro realized that this model required an explanation for why the defeated group would not simply leave and colonize lands elsewhere. He explains that at some point that would not have been possible, since other areas would also be occupied, so "social circumscription" (caused by high population density in adjacent territories) would constrain mobility, and finally coerce the defeated groups to stay. Their land would then be incorporated into the victorious and growing regional polity (Carneiro 1991:207).

According to Carneiro's model, social stratification emerges as the outcome of military conflict, and economic factors do not play an important role in creating power imbalances. However, I believe that although warfare was important in establishing territorial claims—even though evidence for warfare is lacking in the archaeological record—the ability to produce and control resources was critical for the creation of social stratification (Isaac 1975:127).

The most important protein source in Amazonia was fish, but because it was a seasonal resource, fish had to be captured in great amounts and stored. Ethnohistoric sources agree that in Amazonia, fish was consumed fresh, dry, and as fish flour; moreover, thousands of turtles were kept in corrals (de Acuña 1859). As a result, kin groups that had the ability to store animal protein in fisheries would attract other groups that sought stable resources and protection.

The ideological control over the creation and management of these resources—monumental ceremonial mounds close to dams and fish ponds—established "ownership," or the right to manage production.

In assuring an abundance of protein sources through ideological or religious control, the elite could take advantage of a labor force not only to exploit these resources, but also to reinforce their power through the construction of mounds for elite residences, ceremonies, and feasting.

Given the success of new means of exploring and guaranteeing resources, an Expansionist Marajoara period (Schaan 2004) is proposed as a phase during which earthworks and water management started. Teso dos Bichos and other sites are constructed as dams as early as AD 400 (Roosevelt 1991; Simões and Figueiredo 1965).

In the savannas, settlement location was chosen on the basis of resource concentration. The development of a water control technology enabled the intensive exploitation of aquatic resources, which may have supported population growth and the multiplication of similar subsistence systems in several locations where ecological conditions were favorable. This process would have taken a few centuries.

By AD 700, regional polities were fully developed. At the Teso dos Bichos site, the period of main development has been dated to between AD 800 and AD 1100, which is contemporaneous with that of the Guajará mound, at the upper Anajás River (Roosevelt 1991:366–367; Table 5.1). Research at the Casinha and Saparará sites (Schaan 2004; Schaan and Silva 2004), nonmound villages of the same period located along the Anajás River, shows the spread of the mound-builder's cultural traits 75 kilometers downriver. This period, called Classic Marajoara, spans from AD 700 to AD 1100. Much of the information about Marajoara societies produced by previous research is dated to this period.

The typical Classic Marajoara period settlement has the following characteristics: (1) mounds found in groups, usually aligned along the shores of a small stream at river headwaters; (2) one or two mounds out of each group used for funerary rituals, feasting, and elite residence; (3) elite mounds contain retaining walls, baked clay surfaces, baked clay platforms, stoves, hearths, burials, and decorated vessels used for feasting and rituals; and (4) episodes of more intensive mound building, are associated with construction of fish ponds and dams; thin layers of charcoal and baked clay surfaces, and hearths and stoves, are interpreted as occupational surfaces.

Lack of substantial research in nonceremonial mounds prevents a more comprehensive comparison between the two basic mound types (habitation and elite/ceremonial). Mounds containing remains of only domestic vessels were likely habitation sites for the nonelite population, where funerary structures have not been reported. "Habitation" mounds largely outnumber elite/ceremonial ones, as expected in hierarchical settlement systems (Flannery 1976; Johnson 1977).

I propose that the population living on the 34 mounds at Camutins would have numbered about 2,000 people. Since little is known about settlement systems in other parts of Marajó, it is not possible to determine the number of regional polities on the island, making it difficult to estimate the overall population. However, regional aggregations of mounds displaying settlement patterns, cultural features, and material culture similar to those identified at the Camutins site suggest the existence of several small regional polities across the savannas. The greater number of Marajoara phase sites compared with those of previous phases indicates that population growth was associated with the new way of life.

Population growth probably resulted from cultural changes rather than caused them. By AD 700, populations settled along the Camutins River had built a water control system that included dams and large pond excavations. Based on current practices of landscape management and fishing strategies, these earthworks are interpreted as a means to intensively exploit aquatic resources.

The beginning of earthmoving activities on the island is unknown. Current populations living along main rivers seasonally move to the headwaters to harvest fish using simple techniques, such as damming the rivers and placing removable fences and nets in optimal locations. Cooperation between several fishermen, who usually share the product among those who have helped in the construction of fences and dams as well as those who participate in their maintenance and defense, has been documented.

Fishing is far more productive at the headwaters, where water availability is dependent upon constant management. Therefore, collective facilities are preferred, because the work is shared between several households, keeping labor involvement low while productivity is still high. Resource surpluses are high in proportion to labor investment; thus, cooperation is the best way to maximize production and minimizing labor.

Resource surpluses were likely administered and controlled either by those who first started the system, or by whoever was able to claim rights over the fisheries. These groups may have soon invested themselves with the power that abundant protein would give them in the form of a "staple finance" (D'Altroy and Earle 1985), which could also be used to trade for other foodstuffs and goods. Before this system came into existence, economically autonomous households had to cope with resource fluctuations, especially protein-rich sources. As they specialized in capturing aquatic fauna in areas where other resources were scarce, their need for other foodstuffs and goods might have encouraged them to engage more actively in exchange. Although there are no archaeological data besides lithic tools to evaluate exchange systems, ethnohistorical information indicates that trade was an important part of Amazonian economies (Denevan 1996; Lathrap 1973; Whitehead 1994). This might have attracted populations that saw advantages in joining the intensive fishing economy.

Since the elite's rights were justified by a sacred genealogy, productivity in this system may have been credited to the protection offered by ancestors and gods. The elite's close (kin) relationship to ancestors likely sanctioned the relations between humans and natural resources, establishing the rules through which the system should operate to attain maximum productivity and well-being.

In the Camutins polity, settlement patterns, landscape modifications, and the nature of the funerary practices investigated indicate that access to critical resources was regulated by social rank. Only four of the 34 mounds that comprise the Camutins society were located next to the main fishing facilities. Among them, two higher mounds and one small mound contained elaborate burials, suggesting that restricted access to resources was justified through ritual and genealogy, as proposed for other complex societies (Curet and Oliver 1998; McAnany 1995).

The time and resources spent in funerary rituals, as well as the associated iconography (see Chapter 3), point to the importance of genealogy in sustaining social hierarchies. Through funerary rites, which took place in the higher mounds, the elite defined itself in relation to the territory as the land of their ancestors, legitimizing the status quo (Drennan 1995:94–95; Earle 1990:74; Helms 1998; Service 1962).

Mound building also has to be evaluated in light of its symbolic meaning. On a flat landscape, high earthen mounds might be seen as a symbol of power and prestige. The fact that the inhabitants of high mounds are showing their ability to mobilize labor, and mounds can be seen as territorial markers, signaling ownership of land and resources. The kin groups' ability to mobilize people for such an enterprise was likely related to their capacity to manipulate religion by claiming access to the supernatural world.

The monumentality of the mounds can then be understood as a metaphor for social distance and differential access to the supernatural world (McAnany 1995). A comparison can be made to Neolithic megalithic monuments in Europe, which have been interpreted by archaeologists as indicating claims over resources in situations of social stress (Renfrew 1976), and also as symbolic structures that would legitimate sociopolitical strategies (Gilman 1976; Hodder 1982; Shennan 1982).

In Amazonia, there are ethnographic examples of societies where social rank is equated with the right to labor or to zones with particularly abundant resources. For example, among the indigenous communities of the upper Uaupés River, symbolic notions of kinship and social hierarchy determine the distribution of ecological zones; high-status groups are established in areas of aquatic resource concentration (Chernela 1997:279–282).

Since areas occupied by the elite (given their nobility and ties to ancestors) are highly productive, they are expected to be generous, and to promote ceremonies that include feasting and the exchange of food and goods. Commoners are expected to help by preparing for the ceremonies, and the elite's generosity is praised and eventually converted into prestige (Chernela 1997:282).

In the two higher elite/ceremonial mounds at the Camutins polity, numerous remains of ceramics used for food processing along with a large number of decorated vessels represented feasting remains, suggesting that elite-sponsored feasting was used to promote community integration as well as to reinforce social rank (Junker et al. 1994). Feasting likely promoted ceremonial exchange and resource distribution while also reinforcing the rights and obligations among elite members and commoners. Having privileged access to resources, the elite amassed huge surpluses that could be exchanged for other foodstuffs and labor. The elite held control over a stable and desirable protein source, but their circumscription certainly prevented their access to other resources.

This situation likely created interdependence between the elite and commoners as well as among elite groups. Ethnographic research has demonstrated that feasting may be an important arena for competition over basic resources and political power, while at the same time creating mechanisms for a support network in times of subsistence stress (Clarke 2001:144).

The archaeological record at the Camutins site is consistent with this scenario. The relationship between the elite and commoners was likely one of cooperation rather than overt social coercion. Roosevelt (1999) has argued that the absence of any indicators of social coercion would suggest social equality. However, her data point to the existence of social hierarchies and differential access to resources. The absence of social coercion in the archaeological record suggests a relationship between people of unequal social ranks based on mutual obligations. Accordingly, commoners would contribute with labor and possibly other products such as starch, over which the elite likely did not have direct control. The elite would reciprocate with protection and protein resources, due to their "generosity" as well as their special access to the gods. In situations of subsistence stress, leaders may lose confidence whenever they are unable to explain resource fluctuations within the ideological and religious premises that sustained their power (Richardson 1994), which explains the fact that chiefdoms can be cyclical or even collapse suddenly.

Nonagricultural complex societies have often been absent from the discussion on chiefdoms, being labeled as "complex hunter-gatherers" or "complex foragers" (Arnold 1996b; Johnson and Earle 2000:262–263). Some complex societies in the northwestern United States as well as the Calusa in Florida, however, displayed all the characteristics commonly associated with chiefdoms, such as a social hierarchy based on ascribed status, paramount chiefs, a capacity to mobilize labor and request tribute, surplus production, slave labor, competitive feasting, part-time specialization, and warfare (Ames 1995; Arnold 1996a; Marquardt 1992; Widmer 1988). These societies settled in areas of predictable and highly productive aquatic resources where protein could be obtained in greater amounts than usually necessary. Widmer (1988) points out that societies that initially adapt to aquatic resources in highly fertile estuarine zones also tend to develop agriculture. Therefore, in the process, fishing might become a specialized activity within a much more complex economic system. Another possibility is that agricultural products become incorporated into the economy through trade (Widmer 1988:277–278). This incorporation of agriculture might be one of the reasons why chiefdom economies based on aquatic resources are so invisible in the literature.

Common among complex societies whose economies are based on aquatic resources is that resource exploitation starts as a communal activity, but the resources are later appropriated by kin groups that claim ownership of the area and finally establish systems of control over surplus flow and labor (Isaac 1975). Opportunities for such an appropriation may have emerged in the process of organizing resources, storage, and defense. Leadership does not emerge for controlling and distribution resources (after Earle 1977), but administrative roles existed. This might have initially been accomplished by impermanent managers (Widmer 1988:279), but "opportunities" soon would appear (Gilman 1991:148) for certain kin groups to develop more permanent systems of control, or a political economy.

## The Marajoara Demise

The decline of the Marajoara political economy is observed in the abandonment of M-17 (the second in the hierarchy) around AD 1000, the abandonment of the Casinha site (at the periphery of the Marajoara phase domain) ca. AD 1150, and the change in settlement patterns showed by late sites (Cacoal phase, AD 1200–1600) associated with Marajoara phase culture traits (Schaan 1999). The reasons behind this crisis around AD 1100 to AD 1300 are not well understood.

Roosevelt equates this period with the "Teso (mound) subphase," which is not well defined (Roosevelt 1991:242). Agreeing with other researchers (Palmatary 1950), Roosevelt notes that *caraipé*-tempered ceramics would appear in the top deposits, in contrast to the traditional grog-tempered Marajoara phase ceramics. This new technology is observed at the Anajás River sites around AD 600, but only a few sherds can be found at the Camutins site mounds in the top deposits added after AD 1100.

The Leal, Cacoal, and Vista Alegre sites, located along the Anajás River, represent a post-Marajoara phase that provided information on village life after the collapse of the Marajoara culture (Schaan 2004). These occupations span from AD 1300 to AD 1600, which is partially contemporaneous with European contact (beginning of the sixteenth century at the mouth of the Amazon).

Ethnohistorical documents report that Marajó Island was occupied by several different indigenous nations during the seventeenth century, totaling a population of about 40,000 people. Each one of these nations would have a chief, or *cacique*, who ferociously defended their territories (Hemming 1978; Vieira 1992). The archaeological record shows small, apparently autonomous villages lacking ceremonial structures. Ceramic assemblages still include Marajoara-style vessels, but there is greater local variability in vessel shape and decoration as well as manufacturing techniques.

The Cacoal site was 100 meters wide, extending for about 300 meters along the left bank of the Anajás River, some 75 kilometers downriver from the Camutins mound group. Features such as garbage dumps, domestic pottery, and unremarkable ornamented ceramics indicated a small village, perhaps with two houses. However, Excavation 3 produced a number of unusual artifacts, such as 22 tanga fragments, two fragments of spindle whorls, and one cylindrical nephrite bead, recovered from ADE strata I and II. The occurrence of these items only in the lower levels might mean that their usage was discontinued toward the end of the occupation period of the site (Schaan 1999).

In Excavation 4, where geophysical survey indicated an anomaly, we uncovered three small broken vessels. Two of these vessels contained charcoal and ashes, but no bones. Two burials consisted of two incomplete vessels each, one inside the other, and broken pottery. A charcoal sample recovered from the bottom of one of the vessels was dated to AD 1290. The presence of Marajoara-style ceremonial ceramic sherds indicates a continuum of ceremonial life in the Cacoal site following Marajoara culture patterns in the beginning of the occupation.

## Earthmovers across the Amazonian Floodlands

The emergence of regional polities in Amazonia in the first millennium is accompanied by extensive earthmoving activities for resource management, such as raised-field agriculture and an intensification of fishing. Raised fields were identified in the coastal Guyanas, Venezuela, Colombia, and Bolivia. The construction of ponds and weirs has been reported in Bolivia at Llanos de Mojos, on Marajó Island, and possibly in the coastal Guyanas. Raised fields as well as fish weirs were part of a management system for flooded lands, associated with elevated platforms for habitation, burials, and road systems.

Across the Amazonian periphery, characterized by flooded savannas, high savannas, swamps, and river floodplains, landscape transformation started between 500 BC and AD 500, marking the beginnings of the formation of regionally integrated polities based on the intensification of food production systems.

In the Bolivian Amazonian savannas, Denevan (1963) identified and discussed the functions of the various earthworks, such as drained fields, raised fields, settlement mounds, causeways (*calzadas*), canals, and circular ditches. Initially, scholars stressed the earthworks' agricultural functions, but Erickson described water management using fish weirs, ponds, reservoirs, dikes, and canals. By the seventeenth century, the Llanos de Mojos was occupied by many ethnic groups, including the Mojos, Baures, and Cauyuaba, who may have been responsible for these impressive modification of the environment, an achievement that Erickson has labeled "landscape domestication" (2006).

The historical descriptions of the Llanos de Mojos emphasized their farming skills; large settlements with plazas, temples, and streets; roads, canals, and causeways; and regional sociopolitical organization. This explains the reason why, before the Spanish conquest of that territory, the mysterious Mojos fueled the stories about the legendary Paititi so intensively (Gott 1993). I will return to that in Chapter 5, where I shall discuss ring ditches both in Brazil and Bolivia.

The similarities between Marajó Island and the Bolivian lowlands are clear and instructive. Both environments provided plenty of opportunities for intensification of fishing, but landscape management was necessary to create effective production.

Across the Baures region savannas, Erickson (2001) identified zigzag earthworks as permanent structures to channelize and trap migrating fish. Extending for more than 500 square kilometers, these fish weirs were connected to artificial ponds where fish was stored when the waters descended. As at Marajó Island, such strategies provide both a concentrated resource (live fish) for future and easy consumption and the possibility of harvesting high amounts of protein for a growing population.

In addition to building the cultural landscape with the purpose of modifying the environment for food production, other constructions were necessary for living, defense, circulation of peoples and goods, and communal gathering. For this reason, pre-Columbian engineers also built village mounds, elite/ceremonial mounds, moats, canals, dams, defensive and retaining walls, causeways, and roads. Such structures were not exclusively built with dirt. People used several types of trees, palm branches, vines, and other perishable materials to erect houses and temples on top of mounds, bridges to connect mounds and raised fields, palisades to enclose habitation sites, and retaining walls for sustaining dams and platform mounds. Such constructions did not resist the humid climate and the intensive biological activity present in the tropical landscapes; today, they are mostly incorporated into the archaeological record as modified soils.

Earth scientists have shown that high concentrations of some chemical elements in anthropogenic soils are a signature for discarded materials. For instance, zinc and manganese, which are found in high amounts in ADEs, indicate the addition of remains of rotten trees and fibers, likely palms—the main construction material for indigenous houses in addition to their use in the region to produce an array of domestic goods, such as hammocks, baskets, huts, and mats (Kern 1996). The lack of brick or stone buildings, ladders, and pyramids in Amazonia has put the region at a disadvantage to the Andes and Mesoamerica when it comes to discussing social complexity during pre-Columbian times.

Comparisons between the Andes and the Amazonian lowlands often relied on such constructions as achievements attained by a complex interaction between culture and environment. An important factor that has impaired scholars' abilities to understand that ancient Amazonians also built impressive monuments is the low visibility of such constructions for the average observer.

Eventually, earthmoving as a way of managing landscapes and its necessary seasonal maintenance slowed down, completely ceasing at some point during the first two centuries of European colonization. Afterwards, land was either covered by forest and shrubs, or slowly modified through leaching and erosion. Amerindian landscapes studied in the twentieth century were the result of at least three centuries of abandonment, and thus did not retain all the features that were visible when they were in use. This is why, to describe ancient engineering and landscape management practices—some of which survive among indigenous and peasant communities to this day, while others have been abandoned and lost—it is necessary to view landscapes through an archaeological perspective, while keeping a historical ecology–oriented frame of mind.

When studying raised fields, mounds, canals, and ponds, one must keep in mind that Amerindians adapted similar techniques to particular landscapes, and that great variation occurs even inside the same general region. For example, raised-field agriculture in the coastal Guyanas—an area characterized by low swampy plains and seasonally flooded savannas—allowed for the development of at least four different types of elevated gardens: elongated and narrow, rectangular, medium, and small rounded fields, found clustered in large rectangular fields or disposed along the coast in extensions that surpass 100 kilometers (Rostain 2010).

The function of the elevated fields was to control flooding, since a system of adjacent canals would drain the excessive water as well as provide circulation paths between fields (by canoe in the rainy season and by foot during the dry period). Continuous management would consist of depositing a new, fresh layer of silt on top of the fields, furnishing new plants with nutrients. Rostain (2010) believes that populations associated with the Arauquinoid ceramic style built raised-field systems from AD 650 to AD 1650, occupying a territory 600 kilometers long, with a probable demographic density of 50 to 100 inhabitants per square kilometer. In Suriname, raised fields could have been built even earlier (AD 350), and they appear to be associated with the Barrancoid ceramic style (Rostain 2010:345).

In the state of Barinas, in the Venezuelan Llanos (a vast tropical grassland plain), ethnohistorical sources described raised fields for maize cultivation, which were archaeologically identified as related to a late Galván phase (AD 500–1000), and associated with mounds and causeways. In a flooded landscape, ridged fields would allow for two harvests a year, which were very much needed to feed a growing population (Spencer et al. 1994).

Zucchi (1985) notes that Amerindians used natural features such as oxbow lakes and springs as part of a network of canals. Similar strategies are visible on the Marajó savannas, since streambeds and small lakes were dammed or excavated to build aquaculture systems. In the Venezuelan Llanos, Spencer and Redmond (1998) have studied kilometers of *calzadas* (elevated roads), with top surfaces up to 8 meters wide and 1 meter tall. The first hypothesis was that *calzadas* provided a dry path through the inundated landscape, connecting sites across the region (Garson 1980). They would enable the transportation of goods and people necessary for the integration of regional polities, which the first Europeans described as being commanded by powerful chiefs.

However, as a result of a regional survey over a 435-square-kilometer area along the Canaguá River, Spencer and Redmond (1998) were able to demonstrate that there was no correlation between *calzadas* and inundated areas. Moreover, the *calzadas* did not connect all the sites in the region, so their function must have been a different one. Those earthworks relate to a period of demographic growth between AD 550 and AD 1000 called the Late Galván phase, when sites were distributed in a tree-tiered settlement hierarchy.

Calculating population figures for all sites, the authors concluded that although only 42 percent of the sites were connected to the major site, demographic differences between sites indicated that in actuality, two-thirds of the populations were connected. Spencer and Redmond then showed that the influx of people to the regional center was important for regional integration. People attended periodical rituals, and the road system served both to connect the overall population to the regional center and to decrease the power of the secondary centers to maintain regional integration.

Continuing our trip through the northern South American lowlands, we arrive at Colombia, where extensive complexes of ridged fields were built in the 500,000-hectare swampy area of the Mompós Depression between 900 BC and AD 1200 (Parsons and Bowen 1966; Plazas et al. 1993). Here, Sinú culture populations dug canals that were perpendicular to the San Jorge River, extending as far as 20 to 40 kilometers and spaced between 10 and 20 meters apart from each other. The elevated fields between canals received fertile sediments deposited by yearly flooding, whereas the depression between fields remained humid. This canal system extended for 60 kilometers along the San Jorge and other streams, adapting to the meanders of rivers (Plazas et al. 1993), forming impressive designs, like nerves radiating from a leaf stem.

This engineered landscape leaves no doubt about their capacity to adapt nature to culture needs, and thus sustain a substantial population. Along smaller river courses, we find earth platforms for houses with habitable areas that up to 40,000 square meters in size; funerary tombs of various sizes are also present in the extremities of these mounds (Plazas et al. 1993).

In the Amazonian Ecuador, Porras (1987) has described a mound complex along the upper Upano River that was named Sangay, after the nearby Sangay volcano. Porras, who first studied the site, was especially concerned with establishing a chronology and studying the ceramics, so the particular functions of Sangay as a probable ceremonial and political center were not further explored. Salazar (2008) has later developed a more regional and landscape-oriented approach. He describes an interesting type of site found in the province of Morona, Santiago, formed by a group of four rectangular platform mounds around a rectangular sunken plaza (Salazar 2000). Remains of lithic tools and pottery found on the platforms indicate that these places were used for habitation.

Salazar describes sites formed by several groups of platform/sunken plaza units, connected by a road system. Platform sizes range from 50 to 100 meters long, 8 to 10 meters wide, and 2 to 5 meters high (Salazar 2000:40). In some cases platforms are not actually erected; the plazas and paths are dug lower, with platforms formed by the remaining ground. Mound building in the region could have started as early as 500 BC.

Two periods, known as regional development (500 BC to AD 500) and integration (AD 500–1500)—are related to earthmoving practices, which are considered a signature for regional polities, since regional surveys have indicated site hierarchies according to the number of platform/plazas and their heights.

As the studies on Marajó and in other seasonal flooded lowlands across the basin show, extensive earthmoving related to resource management—such as fish farming and crop production—was a recurrent feature in Amazonian societies in the first millennium. They involved the massive transformation of tropical landscapes through managing river courses, controlling irrigation, fish reproduction, and mulching soils, which in turn led to a complete turnover of plant species in those settings over time (Erickson and Balée 2006).

These communal works required coordination and a substantial population. Interestingly, all of the societies cited in this chapter are located in the Amazonian periphery, far from the Amazon River *várzea*, which scholars have historically praised for its bountiful resources. The fact that regional polities appeared first in the periphery is illustrative of the processes that triggered the emergence of complex societies in greater Amazonia, and it highlights the importance of landscape archaeology research as an appropriate approach to the description and study of such societies.

## Notes

1. The expression *Igarapé dos Camutins* has a Tupian origin. *Igarapé* indicates a small river, stream, or creek, whereas *camutin* is a funerary vessel, locally called *igaçaba*. Both words have been incorporated by the local Portuguese dialect. Since *Camutins* has small tributaries itself, it is often called a river or, in Portuguese, *Rio Camutins*.
2. Although both are intended to block the water, and are temporary, there is a difference between *tapagem* and *barragem,* a fact that was also observed by Smith (2002:50). A *barragem* is a small earthen ramp that works well on small watercourses and requires less effort. A *tapagem*, on the other hand, consists of two parallel wooden fences filled with sediment in between. According to Smith (2002:50), the local population uses cecropia (*embaúba*) poles, which rot quickly and are easily swept away by the raising waters in the beginning of the rainy season, allowing for fish migration and the passage of boats.

## Chapter 3

## LAND OF THE ANCESTORS

"COSMOLOGIES AND CEREMONIAL centers that they generate have an absolutely essential role in dissipating the disruptive tensions that human society generates. A loss of faith in the ceremonial life of the community is as fatal to the polity as a loss of sufficient calories to feed its people" (Lathrap 1985:242).

Art is one of the ancient proofs of the symbolic behavior of human beings. The mental development of our species, *Homo sapiens*, is related to the emergence of art and language, differentiating us from other simian anthropoids. Ancient rock paintings dating to the late Pleistocene (35,000–45,000 BC) in Australia, Asia, and Europe are testimonies of ancient cravings for communication that could no longer rely solely on human speech. Art has accompanied us throughout our journey to increasingly complex social relations; the ways we manifest our inner beliefs and desires through visual means of communication has a lot to do with, and is a reflection of, the type of society we live in.

The first Amazonians began painting rock walls and engraving rocks in riverbanks with beautiful designs at least 11,000 years ago (Roosevelt et al. 1996). It was a pleasant way to spend leisure time between hunting, fishing, and gathering nuts and fruits, but more important, it meant the production of visual and permanent landmarks for ideas, campsites, and subsistence territories. However, as societies grew in number and were settled, practicing agriculture, the use of rock walls for expressing thoughts fell into disuse. Other media such as pottery, cotton, basketry, wood, shell, and bone were consequently used. These are portable materials, things that one can carry around, wear, give away as gifts, celebrate important dates with, break, and throw away as one pleased.

In pre-Columbian Amazonia, aesthetic expressions over portable, nonperishable objects became more profuse 2,000 years ago, although ceramic pots decorated with fine incised designs inside broad incision outlined areas indicate careful handicraft taking place as far back as 3,000 years BP (Zone Hachure Horizon style; Meggers and Evans 1961). However, as farming communities multiplied along the basin, settled life provided the basis for improving ceramic production, since clay was abundant everywhere.

Although perishable materials rarely survive in tropical forests because of the high humidity and intense biological activity in the soil, ceramic vessels constitute the bulk of the archaeological assemblages; thus, it is mainly to them that we have to direct our attention when talking about archaeological artifacts in Amazonia. Still, when we look at the decorative patterns in the ceramics, we can guess that some of them were probably an attempt to resemble vegetal fibers, which at times were also used to create the motifs (Torres 1940); the superimposed incised lines found on a number of pottery vessels appear to simulate the designs created by the interwoven fibers used to make baskets, sieves, masks, and so on.

A visitor to an Amazonian village would hardly pay any attention to the ceramics, since so many other items in the material culture might certainly be more persuasive to the eye (Raymond 1995). It is possible, then, that pottery is more important to archaeologists (since it is what has survived for study) than it was to the ancient indigenous societies. However, the fact that sixteenth-century travelers such as Carvajal et al. (2002:61, 86) have mentioned beautiful, colorful, and well-finished ceramic vessels (which were probably polychrome) both in the upper and the lower Amazon regions indicates that at some point, this industry attained a distinctive place in the material life of those societies.

Much of what we know from original Amazonian subsistence systems and collective identities comes from ceramic assemblages. And all we know about symbolic life comes from pottery analysis. This is why this chapter is dedicated both to funerary patterns and ceramic iconography, which will help us to understand the ancient Amazonians' social organization and the particular ways through which they—the Marajoaras in particular—related to their real and mythical landscape, transforming it into the land of their ancestors.

## Depicting Humans

By the time the first regional social formations were emerging in the Amazon estuary ca. AD 400, ceramics had become more sophisticated, with the appearance of the representation of human beings on pottery vessels as well as the production of anthropomorphic figurines. Roosevelt (1988:1) equates that with important changes in the subsistence systems, such as a reliance on an economy based mainly in the cultivation of staple food crops—in opposition to an earlier dependence on hunting— and the transition from rank distinctions to social stratification.

It is likely that, at that point, human representation was related to the increasing power that some individuals, as well as their lineages, were enjoying in society. Indeed, the fact that some persons were buried inside of complex anthropomorphic pottery vessels with varied funerary paraphernalia, including long-distance exchange items, whereas others deserved only plain wares or were buried in the ground, is an important sign of social inequality.

Yet, even though humans are clearly portrayed, these representations are not totally realistic. There is a mixture between human and animal depictions, forming hybrid figures, which is generally understood as a representation of the supernatural (Ribeiro 1987), or a liminal state (Turner 1974). Humans represented in the ceramics are probably ancestors or gods, and not the deceased, since similar images are used in different individual funerary vessels. If individuals were represented in some way, it was by details of which we are not completely aware.

Several anthropologists have emphasized the importance of the "construction of social bodies" (Turner 1980) through ritual in the formation of social identities (Vidal 1992; Viveiros de Castro 1987), as certain rituals are meant to reassure hierarchical relations within society (Heckenberger 1999). In the archaeological record, anthropomorphic images represent chiefs and shamans, seated on stools and bearing prestige items. Heckenberger feels that Amazonian elites were not supported by the possession of wealth, but incorporated power through genealogy, reinforced through frequent rituals (Heckenberger 1996, 2000).

I believe that a more realistic representation of humans—as we see among Andean societies—indicates a lesser emphasis on lineage, with greater importance attributed to individual qualities. In the Amazonian regional polities (with the probable exception of the Tapajó society), the iconography of power was not entirely naturalistic, as was the case among the Aztecs or the Mayas, who left large painted murals displaying diverse scenes of the lives and deaths of powerful people. Conversely, in Amazonia, lineages and individuals sought to preserve their power by leading society to worship their ancestors, who were treated as mythical gods.

In funerary vessels and figurines, humans are usually represented as seated individuals (McEwan 2001) holding emblems of their lineage in the form of animal heads and body parts. One of the Marajó ceramic substyles (Arari Excised), which is characterized by excised decorations over a red-slipped surface, typically depicts a caiman, which was probably the clan's totem. Another substyle (Joanes Painted) features the representation of anthropomorphic figures with mixed owl and female attributes (Schaan 1997).

Found over various areas of different sites, but usually more frequent in cemetery contexts, ornamented pottery was produced for funerary rites, feasts, and other types of ritualistic occasions. Pottery-making, particularly when highly decorated, was part of a prestige system found only in some of the mounds (ceremonial mounds); therefore, it can be assumed that it was also restricted to a few people.

## Worshipping Ancestors

Although plates and bowls with unique decoration and funny shapes found broken in open areas of the mounds are clearly the remains of feasts, most of the ornamented pottery comes from funerary contexts. The funerals themselves were long-lasting ceremonies, which started with the mourning of the dead, and went on to a series of procedures involving the adequate treatment of the deceased until they were "ready" for their final rest.

The corpse was initially wrapped in cloth and put inside a basket, vessel, or even in the ground to wait for its decomposition. After a period of time long enough to allow for the soft parts of the body to be easily removed, the bones were carefully cleaned, painted with red or black dye, and put inside a decorated vessel. This funerary custom, called secondary inhumation, was disseminated among Amazonian societies during the late pre-Columbian period. Anthropomorphic vessels containing disarticulated bones are found both in the upper and the lower Amazon from AD 500 onwards.

Evidently, funerary customs varied from region to region. Cultures such as the Maracás (in Amapá) and the Aruans (in the Caviana and Mexiana Islands, in northern Marajó) did not bury the vessels, but kept them above ground, either in the forest or inside rock shelters (Guapindaia 2001; Meggers and Evans 1957). During the Marajoara IV period (AD 1100–1300), the dead were cremated instead; historical reports on the Tapajó claim that the ashes were mixed with beverages and drunk in special ceremonies (see Chapter 4).

A few characteristics of secondary burials in ethnographic societies have helped us to understand some of the objects and structures found in the archaeological record, as well as to appreciate the importance of such rituals for the reproduction of the social system.

Funerals can be understood as rites of passage, and as such they involve separation from an old structure, a waiting period, and finally, rituals for the aggregation to a new structure and the incorporation of a different social status (Gennep 1977). For indigenous societies, death had to be completed through a process that involved restriction and several ritual procedures, since the person probably needed time to finalize their transition to another state (Hertz 1960).

According to ethnographic sources, Amazonian natives believed that the bones contained the individual soul, and therefore had to be preserved and worshipped (Roosevelt 1991). The long time needed for the decomposition of the body was probably a metaphor for the social and moral transition undergone by each social group (Turner 1967). As life is long, death also had to be stretched to allow for social reordering, since the death of a leader, for instance, might cause social distress (Huntington and Metcalf 1991).

The time the corpse was kept inside a vessel awaiting decomposition might have varied from weeks to months, depending on local climatic conditions. In general, that meant keeping the dead close to the residence, and might have included the periodical removal of fluids, usually accomplished by drainage through a hole made in the bottom of the vessel. Such holes have occasionally been archaeologically identified. A second period started with the opening of the vessel and the preparation of the bones for secondary burial, which may have involved another ceremony and everybody's participation, depending on the social importance of the deceased. Finally, the placement of the vessel inside a shrine, temple, cemetery, or a residential compound called for more feasts and ceremonies, after which life could run its course again.

The types and quality of the goods interred with the deceased tell us about his or her social status. Children buried with expensive jewelry, clothing, or rare objects indicate that they belonged to important families. This is called "ascribed rank," an indication of social stratification that existed in most chiefdoms and ancient states, in which a person's place in the social system was determined by the lineage into which they were born. In Marajoara burials, it is not uncommon to find children buried with rare stone bead necklaces that had to be brought from long distances through trade networks (Schaan 2004:251). Such trade was controlled by chiefs and their lineages, and most of the population had no access to these goods.

When analyzing funerary contexts, it is also important to determine whether differences in mortuary paraphernalia between men and women are simply gender distinctions, or if they implied gender or social inequality. In general, different adornments and objects are placed according to one's individual identity (including gender identity); therefore, differences between genders do not necessarily mean gender inequality. In pre-Columbian Amazonia, emphasis was placed on lineages, as elite persons, independent of gender or age, equally received special treatment at the burial site.

The position of cemeteries relative to other aspects of social life belies the importance the deceased seemed to have, and continue to have, among the living. The relation between burials and habitation sites is important for an understanding of social organization. In most parts of Amazonia, burials were kept close to residences, usually inside a temple or a residential compound. This was also true of Marajó, where burials were found in the same mounds as habitation and feast remains. These mounds are located in the vicinity of the largest fish ponds, which points to the elite's need to live close to their resources and their ancestors. Living on the same grounds as the ancestors was a way of claiming territorial rights and the continuity of privileges, as well as a means of renewing power across generations.

Social inequality was maintained through the active participation of the entire society in the worship of elite ancestors, who were believed to be society's protectors (Curet and Oliver 1998:218; McAnany 1995:160–162). The physical proximity with the deceased was used as a metaphor for cosmological proximity. It was an indication of the connection between a lineage (the living and the ancestors) and a particular territory.

Although funerary vessels were buried in the same mounds inhabited by the elite, funerary and residential spaces were segregated. In all the mounds excavated, urns were found grouped both in the horizontal and the vertical axes. It is common to find superimposed vessels, and sometimes a vessel placed on top of an older one. The custom of always burying in the same place has helped archaeologists detect and study funerary patterns, for whenever a burial is found, others are certainly around.

Excavations at M-17—an elite mound in the Camutins site—have taught us that older burials were primary (i.e., the deceased was immediately placed inside the vessel, without the removal of the flesh that characterizes the secondary burial), since the large vessels containing articulated skeletons were found within a more uniform stratum of silt. In the upper layers, however, the burial method consisted of placing the vessel either on top of a burned clay layer or inside a prepared pit. The neck and mouth of the vessel were kept above ground. This suggests that the burials occurred inside a roofed structure. Several vessels were lidded by inverted plates or bowls, which might have been specially sized for each urn.

An important distinction between funerary vessels that were completely buried and those that were kept at least partially above ground is the amount of access to their contents. Secondary burials generally contain few bone remains, which indicate that the bones were eventually removed from the vessel. In terms of the social meanings of burial practices, it is possible to affirm that unburied vessels permitted closer contact with the ancestors. This may indicate that closely relating to ancestors was important for the majority of Marajoara history.

Ethnographic examples may explain why some funerary urns contained just a few bone remains. For example, Jivaro mortuary practices, reported by Morales-Chocano (2000), included the preparation of the corpse for secondary inhumation either inside or close to the house. Initially, the corpse was allowed to decay; after that, the cleaned bones were placed in a small ceramic vessel that was kept inside the house, close to the roof. Every year, the vessel would be opened, the bones cleaned, the ancestor worshipped, and the bones then returned to their place. Some of the remains would be lost after several years or decades, since damaged bones were discarded. After a certain period of time, only a few bones remained inside a small urn, which eventually was buried under the house floor (Morales-Chocano 2000:86–87).

### Funerary Iconography and Social Identity

Funerary patterns, which involved the type of grave and associated mortuary furniture, are believed to provide information about the social identity of the deceased. These patterns may point to personal characteristics (gender, age, and status) as well as to group membership (Binford 1971). Excavations of Marajoara funerary structures have revealed variations in the size, shape, and decoration of funerary vessels and associated grave paraphernalia both in the horizontal and vertical dimensions, indicating that mortuary practices differed between individuals as well as geographically and chronologically (Derby 1879; Farabee 1921; Ferreira Penna 1877; Meggers and Evans 1957).

Clustered urn burials found in the highest mounds are a sign that only a small part of the total population deserved such special treatment (Figure 3.1). The fact that the burials were clustered in a special area of the mound and that personal wealth was not hugely emphasized suggests that group membership was more important than individual identity.

Researchers have observed geographic variations in the shape, decoration techniques, and designs of funerary vessels within the Marajoara domain. These differences are believed to reflect social boundaries, in

Figure 3.1. Burials at M-17 mound, Camutins site (photographs by Denise Schaan).

the sense that people use style to express group identity (Wiessner 1983; "emblemic style"). On the other hand, it is possible that iconographic symbols on vessels used in ancestor worship ceremonies had ideological and religious meanings. Archaeologists have interpreted artifacts as being actively used by social actors to make statements, to prevent or promote social change, and to advance political goals (Beaudry et al. 1991; Wiessner 1989; Wobst 2000). Earle points out that this "active" component of style is particularly apparent in chiefdoms, where some "elements of style—as in objects used in ceremonial display—are purposefully chosen to signal social relationships and group membership" (1990:73).

In Marajoara funerary vessels, this would be accomplished by demonstrating genealogical ties to ancestors in the iconography to empower the new generations. In this sense, funerary vessel style was likely used to reinforce genealogy and tradition.

Quite a few scholars have excavated and described Marajoara burial customs, myself included. Few nonperishable items were interred with the deceased. Artifacts found in burials included ceramic plates and bowls, tangas, and lithic objects. Meggers and Evans (1957:271–273) excavated a group burial at the Monte Carmelo site (upper Anajás River), identifying the individual in the most sophisticated burial as female; bone remains of two other individuals found in the ground were interpreted as slaves.

During excavations carried out on M-17, also at the Camutins site, we identified an approximately 10-year-old boy buried along with a lithic bead necklace and a lithic axe in a large, undecorated vessel. This was a primary burial, characteristic of the beginning of the cultural sequence for this site, since secondary inhumation later became the rule. The association of long-distance exchange items (lithic beads and axes) with a child has implications for the understanding of funerary practices and social organization. Lithic objects were rare and highly valuable. Only people with high status would be buried with expensive items. The child was probably affiliated with a prestigious kinship group.

Ascribed status is a common feature in chiefdoms, where the social position of an individual is determined by his or her place in the kinship system, and does not depend on personal achievements (Earle 1997; Sahlins 1958). Vessel decoration, in this case, seems not to be important in determining rank, since the vessel was plain. This challenges common assumptions that decoration equates with status. Status is clearly determined by a set of circumstances, which include ceremonies that cannot be observed in the archaeological record.

Some of the observed variations in burial practices were likely related to age and gender, not to rank. Since all the individuals buried in the mounds represent a small proportion of society, it is likely that they were all high-ranking individuals. I believe that vessel shape and iconography reveals information about group membership rather than personal identity. If group membership is more important than individual identity, a uniformity in style is expected within the same site because the living had to demonstrate that they all belonged to the same lineage to show ties to ancestors, and to ultimately justify access to resources.

Snake Iconography

Marajoara iconography is characterized by the use of semirealistic representations in combination with graphic designs. The more natural representations are stylized depictions of humans and animals, sometimes in their entirety, but more often only as body parts. Graphic designs consist of geometric or "abstract" patterns. The meanings of these motifs are not apparent, but they often represent human and animal body parts as well.

One type of design is a three-dimensional representation of humans and animals that might not be completely realistic, but that still maintains basic morphological and structural characteristics. These are, for example, anthropomorphic or zoomorphic head adornments or appliqués, zoomorphic (snake, lizard, caiman) appliqués, anthropomorphic funerary vessels or figurines, bird-shaped vessels, or bat- or turtle-shaped plates, for example (Figure 3.2).

Such realistic representations, however, do not appear in all objects. Most artifacts are more often decorated only with geometric or graphic designs. These designs are used on framing and encircling bands, as

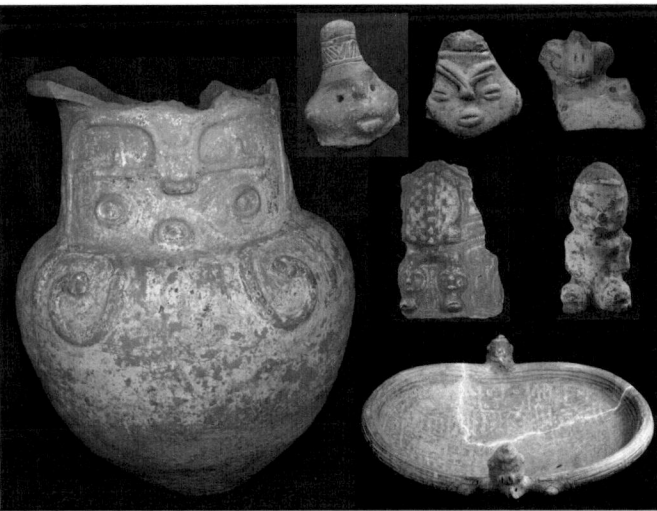

Figure 3.2. Realistic representations on Marajoara pottery (Museu do Marajó collection; photographs by João Ramid).

fillings, as space breakers, or to compose more complex figures over the whole vessel surface (Figure 3.3).

Graphic designs are so profuse, repetitive, and communicative that Ladislau Netto (1885), the first scholar to study Marajoara pottery iconography back in the nineteenth century, thought that they were a form of writing. He went so far as to isolate a number of strokes and compare them to ancient alphabetic writings, finding in fact an impressive correspondence. However, most of the symbols are in actuality pretty common in all human cultures: crosses, spirals, double circles, triangles, and stepped figures can all be found in ancient Greek vessels as well as in most American indigenous art. Therefore, Netto's idea was not pursued further by any other specialist. The idea that the designs conveyed meaning, however, was advanced by subsequent scholars. For example, graphic designs on personal objects, such as on ceramic tangas, were believed to represent a woman's personal identity (Figure 3.7). Nevertheless, the fact that Marajoara designs were ubiquitous on myriad objects supported the idea that they were not more than surface decoration carrying no relevant message.

Anthropologists studying societies who communicate ideas primarily through oral tradition know that "abstract" or "geometric" designs used to decorate objects and bodies are imbued with social meanings, conveying concepts of cosmology and appropriate social roles (Geertz 1983; Otten 1971; Ribeiro 1987; Turner 1980; Velthem 1998). Even though designs might be seen as "abstract" by an outsider, they carry meanings that are shared by the social group, thus being used as a means of communication (Munn 1962, 1973). As Geertz (1983:99) has pointed

Figure 3.3. Pottery decorated with geometric designs (Museu do Marajó collection; photographs by João Ramid).

out, art brings ideas to the concrete world of objects, where they can be seen and touched.

It is possible to access the meanings of the Marajoara graphic designs if we decode them by comparing them to realistic or figurative representations. Several repetitive geometric symbols are depictions of animals that exist or have existed on the island, but the particular way they are represented—as hybrid creatures—suggests that they were likely important mythological creatures.

Realistic and stylized snakes, snakeheads, and snakeskin patterns are the most recurrent figures and elemental designs in Marajoara ceramics. A comparison of naturalistic depictions of snakes and geometric designs shows that certain repetitive patterns found on the pottery are representations of snakes or snakeskin patterns (Schaan 1997, 2001b; Figure 3.4).

One of the characteristics of this type of art, also found in other indigenous cultures, is the dislocation of the elements in the designs, which can be isolated from the whole (Levi-Strauss 1993). For example, sinuous lines represent snake bodies, and they are accompanied by triangles which, although outside the bodies, represent their skin patterns. Another motif consists of interlocked spirals disposed along a band that surrounds the body or neck of vessels, or along the flat rim of plates. The spaces between these stylized snakes are filled by stepped figures, or solid Ls (Figure 3.4).

Figure 3.4.  Representations of snakes and/or snakeskin patterns in Marajoara art.

Another type of representation is the repetition of triangles, diamonds, stepped pyramids, concentric squares, rectangular or triangular volutes, and zigzagged lines on a band. These designs are sometimes used along the short borders of excised plates.

Although this type of geometric decoration is common in pottery (Shepard 1956:295–298), in Marajoara ceramics they represent the snakeskin pattern. This is supported by the fact that the naturalistic snakes represented on some vessels display exactly the same patterns (Schaan 2001c:Figure 4.16–4.17). The endless repetition of elements with the same basic structure is a constant reminder of significant cosmological concepts. The use of snake patterns on vessels and objects used for feasting and rituals suggests that the Marajoara religious symbolism included a snake myth.

Why would snakes be so important in the Marajoara religious symbolism? Snakes are featured in Amazonian myths everywhere, and are still present in contemporary tales and stories among native and nonnative populations (Schaan 2010a). Myths recovered by ethnographers and travelers among indigenous communities have helped ethnologists understand these populations' imaginary and particular relation to the faunal species.

In several Amazonian myths, especially among the Tukano in northwestern Amazonia, studied by Reichel-Dolmatoff (1971) and Janet Chernela (1988, 1997), there is a common association between snakes and fish reproduction. The Tukano live along the upper Uaupés River, and their social organization includes social stratification and differential access to resource zones, in the same way that the distribution of material culture and settlement patterns in Marajoara societies reflect their social stratification.

Among the Marajoaras, it is likely that snake symbols were used as part of a religious system that promoted resource abundance. Because religion was a prerogative of the shamans and the elite—who venerated their ancestors as if they were the ancestors of the whole community—they could have used the snake symbolism to justify social stratification. The ethnographic examples here give an idea of how the snake myth could have been used among the Marajoara populations.

### The Snake People

The myth of the anaconda or *cobra-grande* (big snake), in its many variations, is a recurrent theme in Amazonian mythology (Chernela 1988; Reichel-Dolmatoff 1971; Velthem 1994). The Wayana—an indigenous group who live in northern Brazil, Suriname, and the French Guiana—explain that decorative patterns woven with vegetal fibers to produce baskets were copied from the skin of *Tuluperê*, an immense snake who once terrorized the Wayana and Aparai groups (Velthem 1998). According to the myth, *Tuluperê* periodically left the mountain where it lived and came to the village to eat people. When men tried to kill it, the snake transformed itself into a canoe and engulfed them. Finally, the Wayana organized an expedition of men armed with bows and arrows, who killed *Tuluperê*. When they got close to the dead creature, they observed its beautiful skin, and thus learned the designs they reproduced in their basketry (Velthem 1994:120–121). The myth shows the designs as a gift from the mythical protector.

According to the Desâna, a Tucanoan group living in northwestern Amazonia, snakes are the ancestors of fish; thus, snake tales are frequently related to fish abundance (see Reichel-Dolmatoff 1971:255, 267). One of the variations of this myth explains that the sun created everything on earth, except for people and fish, which were created later. All the tribes were transported to the earth inside the live snake-canoe, which was surrounded by fish. The tribes settled at the headwaters, and the position of each tribe along the river corresponds to their position inside the snake-canoe (Reichel-Dolmatoff 1971:26).

The Wanâna, an indigenous community from the upper Uaupés River, also believe that the location of their settlements as well as hierarchical positions among the sibs is determined by their relation to an ancestral anaconda (Chernela 1997). According to Wanâna mythology, the sibs' seniors emerged from the head of the snake-canoe, which explains their location at the lower river course, a rich resource area. The sibs' juniors emerged from the tail, the reason why they are located upriver. They justify their permanence in the ancestral settlements because this is the only way by which the soul can return (Chernela 1997:293).

We do not have any record of myths told by the Marajoara people, but interestingly, the myth of the great snake has survived on the island throughout the colonial period, and can be found to this day among riverine populations.

An Aruan man allegedly told Alexandre Rodrigues Ferreira, an explorer and early naturalist who navigated across the center of the island in 1783, that his people believed in the beginning, the island did not have its current labyrinth of rivers (Ferreira 1974). Instead, the center of the island was primordially inhabited by uncountable snakes. These snakes had no water, and were in great suffering, the reason why they decided to run to the coast in search of water. Because they were so heavy and large, they left their body imprints on the ground as they moved across the dry land, which turned into sinuous canals. When the rain finally came, the paths the snakes inscribed on the land were filled with rainwater, forming the current sinuous, large rivers in the island (Ferreira 1974).

Dalcídio Jurandir, a Marajó-born writer, also presents the snake theme in one of his novels set in the mid-twentieth century. In *Três Casas e um Rio* (Three Houses and a River), one of the characters is an

Afro-Brazilian woman who mourns the scarcity of food caused by the seasonal drought (Jurandir 1994; Pacheco 2009). She sadly describes the decrease of fish, telling her audience that it all happened when the mother snake was sleeping at the bottom of the river, suddenly woke up as the waters receded and decided to leave for deeper bodies of water. In doing so, she took the fish and all the aquatic life with her, causing much suffering.

Snakes also appear in cowboy tales as well. Beginning in the eighteenth century, the island's economy, previously based on fish, was slowly being replaced by cattle ranching. In a novel by Tocantins (2005), a cow defied even the most skilled cowboys, running to the river when lassoed, diving into the water, and breaking the rope. The next day they would find part of the rope rolled up in the riverbank, transformed into an enormous snake.

A more recent narrative was collected by Teles in an Afro-Brazilian community living at the Gurupá River headwaters in eastern Marajó (Teles and Acevedo-Marin 2009). She interviewed two men who believed that two snakes inhabited the bottom of the river, and that their breath was responsible for the daily tides. They inhaled at high tide, and exhaled during low tide. One day, however, one of the snakes left, which explains why the lakes were getting drier and drier every year, and the fish were getting scarcer. They believed that the snake was frightened away by the cultivation of açaí palms and logging, both recent economic landscape transformations.

As these examples have shown, snakes are an important part of the Marajoara cosmological view, past and present. The survival of snake myths to this day indicates the importance of this ancestral being in a landscape that is highly dependent upon the water cycles, and which is defined by floods and droughts.

## Iconographic Variations in Funerary Vessels

Scholars agree that major geographic distinctions among funerary vessel styles can be observed between vessels from the Anajás River sites (Camutins and Monte Carmelo) and the eastern Lake Arari area (Pacoval site and others). Vessels found at the Camutins and Monte Carmelo sites are more often globular in shape, decorated with red-and-black painted designs over a white-slipped surface. The typical example is the Joanes Painted anthropomorphic vessel (Figure 3.5), which displays appliqués of facial features (eyes, mouth, nose, and ears) on the vertical neck, and painted body parts (arms, uterus, and vagina) on the globular body (Meggers and Evans 1957:Plates 75–76; Palmatary 1950:Plates 93–95; Roosevelt 1991:Figure 1–17; Schaan 2001c:Figures 4.12–4.13). The anthropomorphic features are represented on opposite sides of the vessel in a nearly symmetrical way. The vessel displays a female/owl hybrid

Figure 3.5. (A–B) Joanes Painted funerary vessels (unknown private collection; unknown photographer); (C–D) Replica of a Marajoara vessel from Goeldi Museum by Déo Almeida (Denise Schaan private collection; photographs by Ana Paula Schaan).

figure whose face is depicted on the vessel neck, and the body parts on the vessel body. The eyes are rounded and half-closed, a metaphor for sleep or death; the eyelids are shaped like an owl's; and the eyebrows are linked in a protuberant T. The nose is a small beak, and the mouth is a rounded nubbin. The ears are large and the earlobes rounded, as if a disk had been inserted, from which a woven ear piercing hangs. The body does not display any three-dimensional features; discrete body parts are represented as areas painted in red and black over the body, their positions reminiscent of an owl sitting on a pole. The chest is prominent. The arms are stylized and show the skeleton like an x-ray. The bird-like hands, with fingers pointing down, are paired with the belly, over which a red-painted circle represents the womb. Below this circle, a red vertical rectangle represents the vagina. The remainder of the body is filled with geometric designs, mostly triangles, which in other contexts represent snakeskin patterns.

Most of these vessels have the same female/owl individual front view represented in two profiles. Another intriguing figurative element, placed on both sides of the vessel neck between the ears of the two female figures, depicts a trickster—a small grotesque creature with a human body, but a half-human, half-animal head. These figures vary a lot. Sometimes there is a whole foot or a hand inside the mouth, which "might refer to funerary endocannibalism" (Roosevelt 1991:81). In another example, the creature has an arm pointing up and the other pointing down, perhaps establishing a connection between the earth and the firmament (Schaan 2001c). Like the vessel, the trickster is a liminal being whose appearance and existence occur in transitional situations, such as those that characterize the ritual itself. The vessel transports the deceased soul to the afterlife. The "mother owl" is the perfect vehicle, since the owl is associated with death and the mother figure with birth.

The use of the mother owl vessel to contain disarticulated bones is also a metaphor for the particular function of the owl's digestive system. Owls engulf their entire prey, and later regurgitate their bones and skin. Since native Amazonians believe that the soul lives in the bones, the mother owl is an appropriate metaphor for death and rebirth.

On occasion, a small anthropomorphic being depicted as a "split representation" is found inside the womb or in between the two profiles. It seems to represent a baby, but also a frog, because their members resemble frog legs, and the lower limbs are wide open, recalling either a frog, or a birth-giving position. The association between females and frogs is also widely disseminated in South American Indian mythology (Wassén 1934). The presence of the little creature inside the womb might mean that a mother is represented.

In the same burial area where the mother owl vessels were found, we have identified another type of vessel, also decorated with appliqués and red-on-white painting. These vessels are smaller, with a narrower neck, but have the same overall shape. Decoration is symmetrical on both sides of the vessel, depicting a face on the neck and a "smiling face" surrounded by snakes on the globular body. Human features here are restricted to the eyes and the mouth. The wide, toothy mouth and round open eyes on the vessel body resemble those of *Alice in Wonderland's* Cheshire cat, a trickster who seems to be mocking us. Eyes on the vessel neck are small nubbins surrounded by fillets in the form of a drop, a fish, or a long-tailed scorpion. It is possible that all of them represent scorpions, whose powers would amplify one's vision (Roosevelt 1991:87–88).

Vessels originating from the Pacoval site (and perhaps from other mounds on the eastern savannas), by contrast, have a nearly cylindrical but complex shape, and are decorated with incised lines over a white-slipped surface (Meggers and Evans 1957:Plate 78a; Palmatary 1950:Plate 34; Roosevelt 1991:Figure 1.21a–b; Schaan 2001c:Figure 4.16). Parts of the vessel (such as the rims and appliqués) as well as the larger incised lines are painted red.

Some of these anthropomorphic vessels are composed of two parts: a truncated conoidal base, with insloping walls forming the body; and a slightly globular neck, which forms the head (Figure 3.6). The facial features are stylized appliqués, framed by incised lines. In some of these

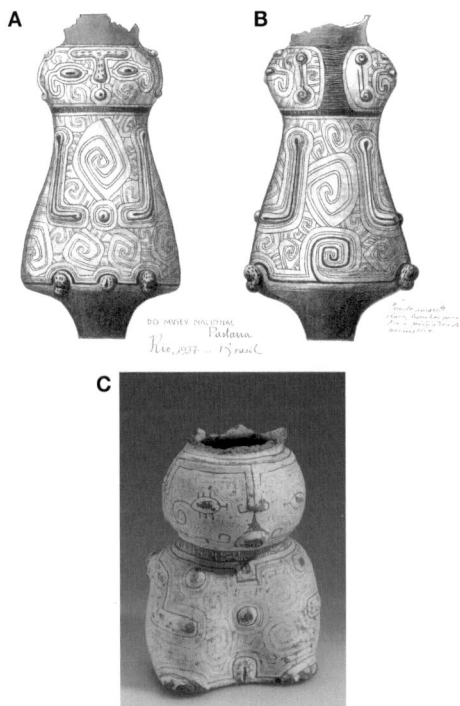

Figure 3.6. Pacoval Incised funerary urns: (A–B) Watercolor drawing of a Rio de Janeiro National Museum vessel, by Pastana (State Museum collection, Belém); (C) Marajoara vessel (AMNH—41.0/1436; Courtesy of the Division of Anthropology, American Museum of Natural History).

vessels, the arms are snakes that originate from a zoomorphic appliqué (sometimes a vulture) at the shoulder level, extending vertically along the body, and curving toward the womb (Figure 3.5), where they might end in a rake-shaped hand (the hand of a caiman). The womb is represented by a circle or a nubbin. The vagina is either a rectangle or a triangle, often filled with punctates representing pubic hair or a tanga. In some of the vessels, three-dimensional but underdeveloped legs indicate a sitting position. The anthropomorphic figure is depicted with a front and a back, unlike the Joanes Painted urn.

This type of vessel iconography brings together snakes, scorpions, caimans, and vultures, showing a preference for powerful, inedible animals to convey meaning in this life's stage. Again, the use of hybrid figures in the burial ritual represents the transformation, the transition taking place, and a need for the support of powerful beings to accomplish the entrance in the afterworld.

There are several similarities between Joanes Painted and Pacoval Incised vessels, such as the representation of a female persona, the position of the hands toward the womb or on the belly, the use of triangles or rectangles to represent the vagina, the use of red paint to emphasize the female sexual characteristics, the use of lines and geometric figures over the body and face (representing body painting), and often snakes.

Joanne Magalis, one of Donald Lathrap's students,[1] studied the iconography of anthropomorphic vessels. Magalis (1975:356–360) believes the two-faced representation in the Joanes Painted vessels (which she calls Tradition 1) and the one-faced representation in the Pacoval Incised vessels (Tradition 2) constitute an important distinction between them in terms of their relation to other Amazonian styles. She considers the Pacoval Incised vessels similar to the Maracá and Aruan funerary vessels, which date from AD 1300 to AD 1600.

Magalis believes that people who arrived on Marajó Island after the Camutins society had already brought the Pacoval style with them. Based on this assumption, Magalis proposed a seriation of Joanes Painted funerary vessels by isolating features that would show resemblance to early or late phases of the Polychrome Tradition, thus providing a chronological framework for the evolution of the style. Although her work is interesting and provocative, she was not able to test it archaeologically, and theoretically it reveals a "passive" view of style (Earle 1990:73).

Although Amazonian styles influenced each other, it is unlikely that these influences were unidirectional. Social groups certainly reproduce styles as a matter of tradition; however, especially in complex societies, certain elements of style may be used or not depending on ideological and political goals. Therefore, the absence of Pacoval traits on some of the Joanes Painted urns may also be seen as a strategy to reinforce social boundaries, instead of the evidence of a lack of contact and of chronological distance.

## Female Objects and Female Symbolism

In prestate societies, the process of centralization of power in the hands of a chief or a lineage is accompanied by ideological justification, which is visible in the material culture and religious symbolism. In this process, both kinship and gender are important in defining social identities and ranks. The iconography of ceremonial and funerary objects often favors anthropomorphic images, conveying concepts on appropriate gender roles. There seems to be a correlation between gender ideology and the degree of political centralization. This is especially clear when we compare the depiction of humans in Marajoara ceramics to that found in other societies.

Tapajó pottery, for example—which I shall discuss in more detail in the next chapter—is very unique and elaborate, almost displaying a baroque style. Anthropomorphic representations appear mainly on medium-size vessels and figurines. Females are usually represented as figurines, with few details. Males are more often represented on vessels, and are usually engaged in some type of activity. Males may have a rattle in their hands, probably representing a shaman.

In Santarém pottery, male activities and ceremonial roles are emphasized. Human representations feature more details than in Marajoara ceramics, and males appear to have a more prominent role. The only anthropomorphic vessels known come from private collections, and there is no record of their contents. We are just beginning to study burial customs among the Tapajó; most authors have suggested that they did not use funerary vessels, and the few that we have found are plain and have yet to be thoroughly studied (Martins et al. 2010).

In Maracá funerary urns, on the other hand, most females and males are represented in a patterned and similar way, as seated individuals who also display animal features. Their bodies and limbs are rounded, resembling a tortoise. There are male and female vessels, inside which human bones were carefully placed. Vessels consist of a cylindrical body lidded by a conical head. These urns are located above the ground, inside caves, all facing the same direction (Guapindaia 2001). Gender differences are specifically indicated by secondary sexual characteristics (vagina and breasts), as well as through the designs and colors painted on the vessel body. Beyond that, Maracá iconography seems to make no important distinctions between genders, which reflects the egalitarian ethos perceived in their settlement patterns (Schaan 2003).

The Marajoara culture differs from the previous two in its greater emphasis on female representations both in figurines and funerary vessels. These representations, however, are not realistic, but a mixture of female sexual characteristics (breasts, wombs, vaginas, and pregnancy) and animal features (of snakes, owls, scorpions, vultures, and caimans). In addition to relying on animal symbolism, the Marajoaras' supernatural world was deeply embedded in feminine principles. I believe that the Marajoara art reproduced the archetypical figure of the Great Mother. It is likely that female deities were important, and that the elite dynastic groups followed a female line of descent (Roosevelt 1988).

Out of all Marajoara artifacts, tangas and figurines (along with the previously mentioned funerary urns) are the most representative of their gender iconography. I shall examine these in some detail.

## Tanga: An Intriguing Piece of the Female Wardrobe

Charles Frederick Hartt (1876), a Canadian-American geologist, was the first to mention the existence of tangas, even before he realized their function. Because he encountered only pieces of them at first, Hartt initially thought that they were spoon parts because of their curvature.

Later, when he found entire pieces, and saw a triangular shape depicted on an anthropomorphic funerary vessel, it became clear that they were aprons. By ethnographic analogy, they were called tangas: female pubic covers similar to those made of fibers or textiles, which are still worn by indigenous females today.

An analysis of ceramic tangas tells us that Marajoara women and girls wore them on at least a few occasions. Some of them may have been worn during rites of passage. Ceramic pubic covers are not used by Amazonian peoples today; the only ethnographic reference to their use comes from a short description of a ritual performed by a Panoan group living in the Peruvian jungle in the beginning of the twentieth century. The account was reproduced by Steward and Métraux in the *Handbook of South American Indians*:

> Tessmann (1928, 1930) reports that a group of pubescent Chama girls were assembled at full moon, while men and women danced all night, and drank from special zoomorphic vessels. The next morning, while stupefied with drink, the girls were painted, and then each one was laid on a stool, where an older woman cut off and buried the clitoris and the labia. Reich (1903) adds that each girl was deflowered with a clay penis representing her fiancé. The girl was then isolated in a hut for a month, wearing an egg-shaped piece of pottery as a pubic cover (Steward and Métraux 1948:583).

Besides the tangas, all the material objects mentioned in this narrative—stools, zoomorphic pots, clay penises—can be found in Marajoara pottery archaeological collections.

Tangas are triangular, convex pieces made from reddish, nontempered clay. They can be red (Figure 3.7B), white, or tan slipped (Figure 3.7A). Red tangas were intensively polished, whereas the white and tan ones received beautiful red-painted designs. Some specimens of undecorated tan or red tangas, however, have occasionally been found inside beautifully decorated vessels. The final color of each piece depends on the type of clay and dyes used, which explains the variety of color hues in museum assemblages.

Hartt explained that strings tied to perforations found on the three corners of the tangas were used to accommodate the piece to the female body. In an earlier study I conducted on a sample of 252 tanga fragments, I examined the grooves produced by the friction between the string and the exterior surface, and concluded that 92 percent of painted tangas and 98 percent of red tangas were worn on the body. Moreover, breakage patterns show that the pieces were consistently fractured at the threaded orifices. That means that the force applied by string pulling eventually caused the piece to break at the edges.

Observing tanga designs, Hartt (1876:28) wrote: "The upper decorative field is almost always filled with a series of colorful triangles. Below

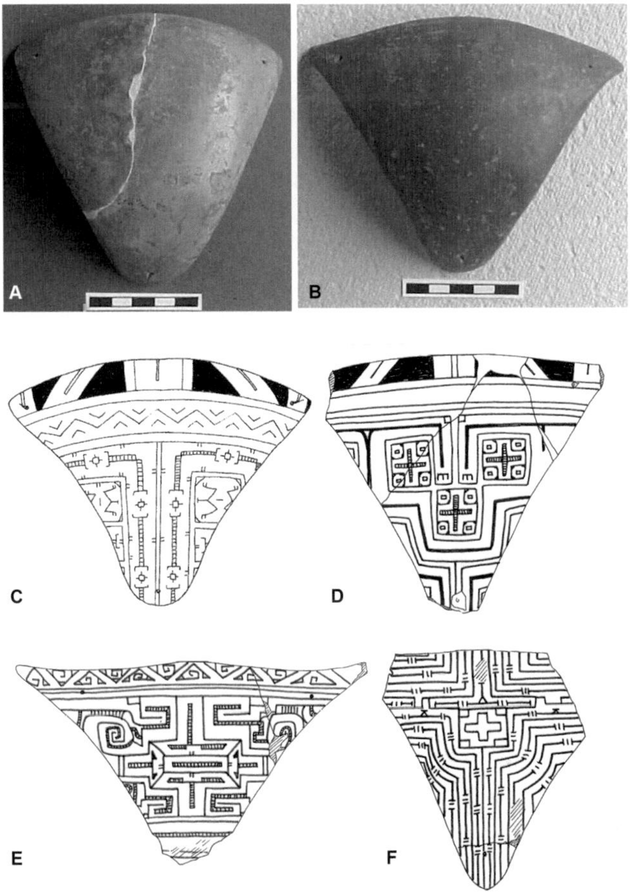

Figure 3.7. Tanga iconography (photographs and drawings by Denise Schaan from Goeldi Museum objects).

this horizontal line, one finds a band formed by two parallel lines, where there are a series of zig-zags or adornments in the form of 'X's, containing other small designs between them. Below this line, there are other intricate, delicate designs."

These decorations, however, were perceived by Hartt (1876) as somewhat incomplete, since the designs seemed to exceed the limits of the object, not respecting its borders. He believed that a possible explanation for this would be an adaptation of designs found on rounded vessels. Another explanation would be that the designs were produced by the suggestion of a shaman during a hallucinogenic trance, with the vision projected over the object as a slide.

Like Hartt, Ferreira Penna (1877) stressed the variability of decorative designs, claiming that tangas would only be worn in particular occasions by married women or those who were ready for marriage. Ladislau Netto (1885) was especially interested in the decorative designs, and identified six to eight different patterns. He interpreted them as different gradations, which would indicate rank orders, marking status differences. He suggested that tangas were used during menstruation, and were probably symbolic of a female deity. According to him, red-painted tangas with no designs would belong to women of low rank. Meggers and Evans reached the same conclusion 50 years later, believing that undecorated tangas would be for daily use, whereas ornamented ones would be worn during ceremonies. Besides that, they noticed a larger curvature angle in decorated tangas compared with the red ones. The range of variation would correlate with differences in the anatomy of the wearers (Meggers and Evans 1957:382). That might indicate, for instance, that red-painted tangas belonged to large women.

Several other authors wrote about tangas during the twentieth century. Angione Costa (1941), for example, felt that tangas were only used during ceremonies, since it would be uncomfortable to wear them on a daily basis. Mordini (1929) stated that most of the tangas were in fact used, since the strings left marks on the surface surrounding the holes. He stressed, however, that some tangas did not have the necessary curvature to be worn, so these would have been used only to decorate funerary vessels.

Based on cross-cultural information on the use of pubic covers, Roosevelt (1991:84) proposed that tangas were used during female rites of passage and, as gifts, they would symbolize "a woman's wifely role."

Whenever tangas were found inside funerary vessels, the skeletal remains were either unidentifiable or female (Meggers and Evans 1957:384). Tangas were found solely in elite habitation and ceremonial contexts, where they make up 4 percent of the total number of fragments recovered during excavations.

Even among the elite, tanga use was restricted; in the Camutins site, tangas were only found in two of the mounds (M-1 and M-17). At those sites, a large proportion of the artifacts were recovered from open areas and garbage fills, relating to activities such as rituals, feasts, and pottery production (Schaan 2004:344–345).

Contrary to most authors' assumptions that ornamented tangas were worn by elite women and plain ones belonged to low-ranking women, the fact is that undecorated tangas were worn by high-ranking women, since they were found buried in beautiful funerary vessels along with several other personal items. As ethnography suggests, it is more likely that decorated tangas, which featured intricate designs and were found in many social and domestic contexts, were worn by youngsters during puberty rites.

As other scholars have already noted, tangas have three main decorative fields (Figure 3.7):

(1) A top horizontal band, 1 to 2 centimeters wide, is ornamented with triangles and rectangles in a pattern (Mordini 1936:62–63). The central icon is a vertical rectangle with a central painted groove that may represent the female sexual organ, since similar geometric representation is found on the pubis of female figurines.
(2) A second, intermediate band, just below the first one, displays a variety of geometric designs interpreted as snakeskin patterns. These are repetitive motifs composed of triangles and diamonds. Similar designs represent the skin pattern of realistic snakes in Marajoara pottery. In addition, they appear in several other objects, often in a prominent position on vessel necks, plate borders, etc. Since there is a degree of variation among tangas in this second band, although there is also a certain regularity, it is possible that, as Meggers and Evans (1957:384) suggested, this band represents a social or religious system in which women belonging to the same group would wear the same motifs. In some tangas, however, the iconic representation of a caiman instead of the snake motif is present in the second decorative field. Both snakes and caimans are also represented in funerary vessels, which might indicate that they were clan symbols. The caiman, as is the snake, might be an ancestry category.
(3) A third decorative field makes up the remainder of the decorated surface, where a considerable variety of designs are disposed in a harmonious way. The greater variation in this field can be interpreted as indicating individual identity.

Taking into account the three levels of information present in tangas, it is possible to envision three levels of meaning. The first one relates to puberty; the second, to the daughters of the ancestral snake (the snake/caiman people); and the third would relate to kinship as well as to other relevant social identities.

The variability found in the sample, however, is more challenging, since the three decorative fields just described, although more prevalent, are not universal. For instance, in some of the tangas, only fields 1 and 3 are present (Figure 3.7D); in others, only fields 2 and 3 are present (Figure 3.7E); and in yet another group, only field 3 is present (Figure 3.7F).

What could explain such discrepancies? On the one hand, they can be interpreted as geographic/ethnic particularities, since in one case, five tangas found at the same site show a similar style (Boulhosa, personal communication, 2001). On the other hand, these variations might represent different social status or distinct moments along the ritual process.

Earlier, I quoted an ethnographic account of a puberty rite performed at menarche, which marks a girl's transition from infancy to adolescence. In this case, social puberty is promoted to conform to physiological puberty, following certain patterns observed for rites of passage. The idea behind this custom is that an old social identity has to die ritually so that a new social identity can be born.

When transitioning to adulthood, adolescents may be subjected to hardships as a group, which also helps foster a bond between them. Their reincorporation into society may be marked by changes in the body, such as tattoos, scarification, new clothing, adornments, and so on (Viveiros de Castro 1987). I believe that tangas would have been appropriate clothing to build social bodies in such rituals.

Similar tanga designs might have represented shared, social identities, whereas designs unique to one specimen might have represented individual identities in kinship and social roles. I feel that tanga ornamentation is as a way to encode a three-layered social status, from a more inclusive to a more exclusive sense of belonging. In this sense, tangas would be a diacritic sign used to emphasize frontiers, indicating group inclusion or exclusion (Barth 1969).

## Interpreting Figurines

Figurines are among some of the oldest specimens of figurative, portable art, produced by humans as far back as 30,000 years ago. Beginning in the nineteenth century, figurines have been found virtually all over the world, generally depicting voluptuous women, which has led some scholars to believe that they were representations of an original pan-European goddess (Ucko 1968). This understanding was supported by Johann Bachofen's ideas on the existence of a prehistoric matriarchy, and a cult to a female deity (Russell 1998). However, the great variability of figurines found, as well as the fact that they have appeared in areas that were both geographically distant and diverse, has led to a review of these earlier hypotheses. Today, there is a vast literature on female figurines. Different reasons for their production and use have been proposed, given the distinct contexts in which they have been found (DeBoer 1998; Schaan 2001a).

In Marajoara sites, figurines are found in elite mounds, both in funerary and nonfunerary contexts. The bulk of the assemblages in museums today, however, come from private collections without reliable information on provenience.

The most intriguing characteristic of these female figurines is the fact that they are shaped like a phallus, thus blending female and male sexual characteristics in the same object. The phallus head is the figurine head; the testicles are the folded legs in a sitting position. Some of the figurines are hollow, and could have been used as rattles. The female characteristics in these objects are the representation of a vagina (as a triangle, a vertical rectangle, or simply a vertical incision), a pregnant belly (in a few cases), and breasts. Some of the figurines, however, do not have any other indication of sex, which I have classified as sexless (Figure 3.8). There is no known case of a male figurine.

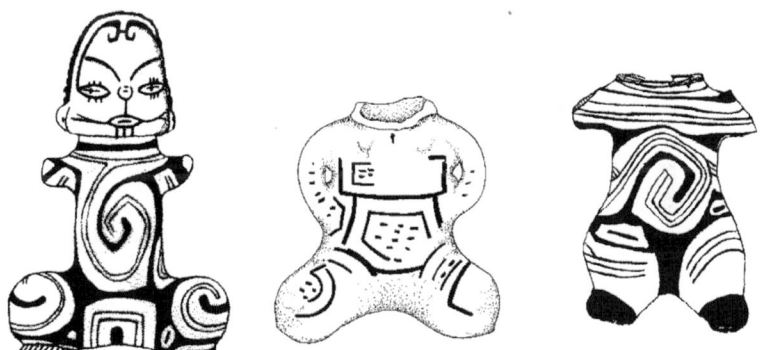

Figure 3.8. Figurines: Styles and patterns of breakage (photographs by João Ramid, Museu do Marajó collection; drawings by Denise P. Schaan, Museu Goeldi collection).

In museum collections, a large number of figurines are found broken at the neck, resulting in a great number of bodies and heads; Netto (1885) called them "idol heads." Out of a sample of 48 figurines I have studied, 33 percent were complete, 31 percent were either heads or bodies, and 10 percent were broken at the waist. My study has showed that 85 percent of the figurines were phallic in shape, 40 percent of which were female. Figurines found broken at the neck were either female or sexless. The neck breakage pattern may indicate intentionality, since the figurines have a robust constitution, and the neck is not especially fragile (Schaan 2001a).

One possible explanation for the breakage could be that figurines were deliberately broken after use. A study conducted by Reichel-Dolmatoff (1961) among the Choco and Cuna Indians along the Colombian coast demonstrated that figurines were used as spirit bearers in healing rituals, and would be broken after the ritual was finished to free the spirit. In this context, where figurines were used mainly by shamans as bearers of spirit helpers, Reichel-Dolmatoff refutes the "fertility theory." He argues that the shaman could use pregnant figurines to protect women during pregnancy and delivery (Reichel-Dolmatoff 1961:237–238).

Marajoara art shows us how the ancient inhabitants of the island "represented themselves and the world" (Roosevelt 1991:76). Given its association with ceremonial and funerary contexts, it has been characterized as religious or sacred, and can be compared with the arts of other Amerindian societies for which we have ethnohistorical or ethnographic information. The profuse representation of humans and animal beings in hybrid or liminal states indicates that the ceramics conveyed important cosmological concepts, helping to provide social cohesion and well-being. The study of the ceramics, as I have intended to demonstrate in this chapter, helps us to grasp how the Marajoara populations humanized nature, and consequently the landscape, which they transformed in the land of their ancestors.

## Note

1. Donald Lathrap (1927–1990; Ph.D., University of Illinois, Urbana-Champaign 1962) was an anthropologist with strong interests in Amazonian archaeology and ethnology. He was an advisor to important Amazonists such as Thomas Myers, Warren DeBoer, and José Brochado among others. His most famous work is *The Upper Amazon,* published in 1970.

*Chapter 4*

# Ponds, Lakes, and Feasts: The Cultural Geography of Anthropogenic Soils

## At the Province of Saint John

WHEN THE FLEET led by Francisco de Orellana in 1542 approached the lower course of the Amazon River and turned an abrupt river bend, they saw many fairly densely populated villages that, they would later learn, were under the dominion of the "Amazons" (warrior women). Reports from that encounter have provided details that led scholars to believe that was the Nhamundá River mouth (Porro 1996:58). The description of this unexpected encounter is as epic as it is tragic. Reportedly, some of the local inhabitants welcomed the Spaniards in "no friendly mood," which would have led Orellana, the captain, to give orders "to shoot at them with the crossbows and arquebuses" (Carvajal 1934:212–213). The natives answered with a long and deadly shower of poisoned arrows, which left many of their men hurt, including Carvajal, the expedition chronicler. The arrow severely injured one of his eyes, which may have impaired his ability to judge their enemy, as Carvajal would later report the attackers' leaders as being 10 to 12 white women. He described them as tall, white, stronger than the men, and very skillful in managing their bows and arrows.

In the subsequent centuries, the indigenous peoples who lived between the Nhamundá-Trombetas and Tapajós Rivers would always be remembered for their poisoned arrows, which caused other native nations to fear them, also earning them the respect of the Europeans. During this first battle, Orellana supposedly captured a native man who would serve as an interpreter during the remainder of the trip. Carvajal tells us that when the captain asked him about the leader who attacked them, the man answered that his name was Couynco. He would also have described the Amazons, women without husbands who lived a seven-day journey from the shore, but who also controlled all the land on the left bank of the river. Their ruling mistress was known as Coñori (Carvajal 2008:220–221). Similar names appear in other chronicles, but the people who inhabited the lower courses of the Nhamundá and Trombetas Rivers would eventually become known as the Konduri.

The supposed relationship between the legendary Amazons and the Konduri survived throughout the colonial period. If at first the Amazons were believed to rule over the Konduri peoples living along the *várzea*, later they would mainly be referred to as the ones who produced the greenstone pendants and beads that both the Konduri and the Tapajó traded in the basin. I will return to this later.

As Orellana and his companions continued their trip down the Amazon River, they were again attacked by natives, who this time came from the right riverbank. The Spaniards decided then to avoid the attackers, sailing along less populated shores. They were told that the rulers of the peoples living on the right riverbank were called Arripuna, Tinamostón, and finally Ichipayo, in this order. This last name appears in different forms in the chronicles, as Chipayo, Rapajosos, and finally Tapajó, which became both the name of the river and of the people living there (Carvajal 2008:223–226).

The Spaniards called this vast region stretching along the banks of the Amazon River—which included the highly populated islands between the Nhamundá and the Tapajós River mouths—the Province of Saint John, since they arrived there on the June 24, 1542, St. John's Day. The islands were described as having high bluffs and an apparently fertile soil; the largest island would have been 18 miles long, which is supposedly the island located between the cities of Alenquer and Santarém (Carvajal 2008:75).

Setting aside the controversial and improbable encounter with the Amazons, which has created skepticism of Carvajal's account, the landscape he briefly described fits well into the archaeological record. Carvajal used the word "province" to designate the settlements that extended along the Amazon River from the lower Nhamundá to the Tapajó Rivers, adding that territories under the same ruler spread well into the hinterland. The term "province" was also used by other explorers to denote populations who shared the same cultural attributes within some sort of political unit, as they clearly indicated that the extension of these provinces "was coincident with their lord's jurisdiction" (Porro 1994:86). Indeed, all of the sites in that region contain archaeological ceramic deposits belonging to the Incised Punctate Tradition (Meggers and Evans 1961), which can be found over a vast region that encompasses the Orinoco and Amazon basins as well as the British and Brazilian Guyanas. Besides their similar material culture, dozens of archaeological sites in the region, some of them especially large, are characterized by deposits of ADE, which are always located in higher lands well above the flood line.

Although scholars have mentioned and studied these sites since the nineteenth century, focusing either on locating them or studying their ceramics and soils, no attempts have yet been made to understand, with the same perspective, the settlements located along the Nhamundá-Trombetas and the Tapajós River basins, which I propose to do in this chapter. I think this is important because settlement patterns in this area show many similarities. I believe that to understand those occupations, we have much to gain from aggregating data that has been produced and presented separately in one analytical effort.

Although no ethnohistorical data are available for the lifestyle of the peoples living on Marajó, travelers and missionaries have provided some information on the Tapajó and Konduri societies, also mentioning others who would have lived in the vicinities. Despite the profound bias and ambiguity that underlie these accounts—especially among missionaries, who believed that the indigenous religious and social practices were inspired by demons—they are worth exploring, as they provide some guidance for building hypotheses that can be tested against the archaeological record.

## The Tapajó and Their Neighbors

After Orellana's voyage in 1541–42, almost a century went by before another expedition followed in his steps, providing information on the peoples who lived along the lower Amazon River. What happened in the meantime? According to Porro (1994), who carefully examined and compared the sixteenth-century chronicles (Carvajal, Vasquez, and Altamirano) with those of the second period, beginning in the seventeenth century (de Acuña, Heriarte, and Laureano de La Cruz), important population losses occurred during this period, even if we consider that European contact was still sporadic and probably not intrusive at that time.

The main factor for the rapid population decline that occurred in the first years of contact was likely the spread of illness: the European intruders carried with them diseases that the Native Americans' immune systems were not prepared to fight (Wolf 1997:133). A second factor would have been the consequences of economic changes. Throughout the sixteenth century, the use of indigenous trade networks by the European trade companies, leading to both the introduction of metal tools and a demand for raw materials, would gradually destabilize the indigenous trade systems. These systems were intermeshed with warfare and raiding, also involving the control of human labor and territories (Whitehead 1994). The changes brought about by the introduction of new products and demands would affect the indigenous political economies greatly, causing power imbalances, changes in the existing subsistence systems, and ultimately cultural change. Finally, slavery was also responsible for the huge population decline that followed the first two centuries of contact.

During the first decades after Cabral's arrival in northeastern Brazil, the Portuguese Crown was particularly interested in the eastern coast.

They soon realized the existence of well-established European companies in several parts of the lower Amazon (Edmundson 1920:54; Guzmán 2008). Although many native groups traded and made temporary alliances with the foreigners, the relationship with the Portuguese was less friendly. The Portuguese raids for slaves as well as punitive expeditions were responsible for the indigenous peoples' eventual attacks on merchants and colonists (Chambouleyron 2008; Guzmán 2008). For example, missionaries mentioned the account of an English ship that brought colonists in an attempt to settle and to plant tobacco along the lower Tapajós River, only to have their crew attacked and killed by the natives ashore (Bettendorf 1990; de Acuña 1859).

Sensing that European trade companies were essentially seizing the Amazon River, the Portuguese initiated a reaction against them in 1616, when, after the expulsion of the French, a military expedition left Maranhão for the mouth of the Amazon River. As they mistakenly took the Pará River for the Amazon, they ended up building a fort in the mainland, across from Marajó Bay. Between 1622 and 1629, a series of battles between the Portuguese and the Dutch resulted in the extinction of various foreign trading posts as well as the destruction of the forts built by the Dutch on Gurupá Island and along the Xingu River. The struggle for the control of the Amazon River would last 31 years, during which the Europeans made several alliances with the native peoples, who would fight alongside whoever offered them more advantages (A. da S. Lima 2006).

During the war against the Dutch, Pedro Teixeira played an important role in their defeat (Edmundson 1920) and sailed up the Tapajós River for the first time, becoming the first Portuguese to explore it. In 1626, when entering the Tapajós River, Teixeira visited a village that appears to have been Alter do Chão, where the Jesuits later established a mission, calling it Borary Village. Teixeira mentioned an abundance of food as well as an impressively large population.

Teixeira would also be the first to travel up the Amazon River in the seventeenth century, motivated by two Franciscan friars who had successfully arrived at the mouth of the river after leaving Quito in 1636. Teixeira took Domingo de Brivea (one of the friars) with him, along with a large fleet of about 1,500 people. His trip back from Quito was described by Cristóbal de Acuña, a Spanish Jesuit father who wrote one of the most reliable reports on the people who inhabited the river, their customs, and locations (de Acuña 1859). Although the Tapajó and the Portuguese enjoyed a friendly relationship during the first half of the seventeenth century, Bento Maciel Parente, the state governor's son, managed to convince his father that the natives were preparing a rebellion to attack them and capture slaves (Nimuendajú 2004:118). The Tapajó suffered a great defeat in this war, which took place in 1639.

According to historical sources, Parente requested 1,000 slaves from the native population after winning the war, but the Tapajó reportedly had only 200, so they had to deliver their own relatives as slaves. In subsequent years, the Portuguese would repeatedly request slaves, and the Tapajó, having no slaves to deliver, seemingly pointed to neighboring tribes as runaway slaves, asking the Portuguese for help in capturing them (Bettendorf 1990).

Given the violence inflicted by the Portuguese, the Tapajó's attitude in the following years would be that of unconditional submission. When Teixeira returned from his voyage to Quito in 1639, he was welcomed with abundant food and beverages. This year marks the beginning of the end of the Tapajó as an autonomous group.

A report by Friar Laureano de la Cruz on the Franciscan friars' 1651 voyage from Quito to Pará makes it clear that huge population losses had already taken place. He mentions arriving at a place with six houses, inhabited by the "Condurises," which was also the name of the river. One hundred and fifty miles below they would have found the "Rapajosos," who inhabited a 10-house village. He describes the trade of slaves for goods: one slave was worth three tools, a shirt, and two knives (Cruz 1900:121). In spite of the pressure on the part of the Portuguese for slaves, it is possible that the Tapajó themselves were in fact accustomed to having slaves, which were probably war captives.

In 1661, Father Antonio Vieira sent Friar João Felipe Bettendorf (1990) to the Tapajós River mouth with the task of establishing a mission there. Contemporary sources mention the existence of two main groups in the area: the Tapajó and the Urucurus (the latter inhabiting the left riverbank), who supposedly spoke different but related languages. It is clear, however, that neither of them was a Tupian language, since the missionaries did not understand them, and although the catechism was allegedly translated into both languages, they left no written record of them. Under Jesuit rule, the abuses against the natives worsened, since the friars compelled them to change their customs and abandon their traditional religious life.

It is possible to infer a few things about the groups' sociopolitical organization from Bettendorf's accounts. Upon his arrival, five principals welcomed him, bringing tortoises and fruits. The next day, other chiefs also came from the hinterland. Although lacking in detail, Bettendorf reported that these populations lived in multifamily houses accommodating 20 to 30 people who were headed by a local chief, and that all the family chiefs were under the rule of a regional lord. This structure is in accordance with a chiefdom configuration, in which there would be a chief above the village level (Oberg 1973 [1955]).

Even with population decimation, the Tapajó represented the largest population along the Amazon River at the time of Bettendorf's stay among them. It is possible that many natives had already fled for the middle and upper river courses at this point, trying to escape both the constant slaves raids and the rapid spread of diseases (Heriarte 1874 [1662]).

Bettendorf's reports of his short stay among the Tapajó are filled with episodes of ethnic violence against the local population. The Tapajó were polygamous, which the friar considered a great heresy. As a result, Bettendorf wished to persuade them to marry only one wife. He eventually came up against the issue of social stratification among the Tapajó when a prospective husband could not be found for one of the women, Moaçara, who was part of a noble family and could only marry someone who belonged to an equally prestigious social stratum (Bettendorf 1990).

Bettendorf mentions that the native people adored the mummified remains of their ancestors, which they kept in sanctuaries and worshipped with dances and offerings. One of those mummies, called the First Father, was kept in a chest next to a house roof. The mummies were set on fire under Bettendorf's orders, but the Tapajó did not react to this abuse, as they were afraid of retaliation. According to later accounts, the custom of mummifying important individuals continued, but the bodies were then secretly maintained in houses in the forest, away from the missionaries (Daniel 1840).

The local population was also forbidden from meeting in plazas for chanting and drinking, and whenever they were caught doing so, their beverage containers would be broken. These celebrations were briefly described before being banned. According to Heriarte,

> ... they have painted idols that they adore, and to which they give a tenth of the maize they produce. ... They say that this is "Potaba de Aura," which in their language is the name of the devil; from this maize, they make some wine every week, and on Thursday night they take it in big vessels to an open yard behind the village, which they keep very clean, where they all get together, and play mournful and sad songs with trumpets and drums for an hour, until an earthquake takes place, and then the devil comes and enters an enclosure built for him, so they dance, and sing, and drink, until it is over (Heriarte 1874:36).

Women had to bring the wine to this "devil's yard," but upon arriving, they would have to crouch down and cover their eyes with their hands so they would not see (Bettendorf 1990). Interestingly, this is the position of the small female figurines in a typical Santarém phase vessel (Gomes 2001; Palmatary 1939), found in archaeological contexts, but not mentioned in the chronicles (Figure 4.6A).

Bettendorf left the Tapajó in 1661, when the Jesuits were banned from Brazil for the first time. The information he left on the Tapajó way of life was complemented by a report written in 1662 by Mauricio de

Heriarte, a high-ranking government official in Maranhão at the time. His report provided data on the Tapajó's subsistence system, emphasizing maize cultivation, with manioc a less important crop. Heriarte acknowledged, however, that this did not conform to the general trend in the basin, since most indigenous populations relied heavily on manioc for calories. Tapajó beverages, however, which according to the friars were the most important element at their feasts, would have been produced mainly from maize.

Thirty years later, in 1691, Bettendorf returned to Santarém to find the village totally ruined. His mission had been replaced by a fort, and the natives had been relocated to other settlements.

Another mission would be founded along the left bank of the Tapajós River in 1698, with the Arapiuns and probably a few Tapajó, in a place later named Vila Franca. Surveys conducted in the 1920s by Curt Nimuendajú (2004) along the Arapiuns River and on the south margin of the Vila Franca Lake revealed numerous archaeological sites, which suggests that these places were indigenous villages that existed previously to the establishment of the mission. According to Nimuendajú (1949), the Tapajó and the Urucuru no longer existed at that point.

In 1835 and 1836, native peoples living in the village of Alter do Chão, the former Aldeia Borary, attacked the city of Santarém, but were defeated by the Portuguese and their allies, the Munduruku, who were moving from the Madeira to the Tapajó River. When Henry Bates visited the region in 1852, he found no elder men living in the village, and the people were in complete poverty (Bates 1892).

When Barbosa Rodrigues visited Santarém in 1872, the city had only about 2,300 inhabitants, since the population had greatly decreased in 1855 due to a cholera outbreak. The disease was devastating among Tapajó descendants and other indigenous groups, who still lived in Aldeia Borary (the original village, now one of the city districts).

## The Konduri

There is much less historical information on the Konduri. The Trombetas River (Trumpets River), an allusion to the indigenous custom of playing trumpets very loudly in times of war and during celebrations, is known to have been home to several different indigenous nations (the Bobuis, Aroases, Tabaos, and Curiates, etc.), among which the Konduri was the most famous (Heriarte 1874 [1662]). The name Konduri, used with some variations by all chroniclers, was first penned by Carvajal: a native man told him that the supreme chief of the whole Nhamundá region was a woman named Coñori (Carvajal 1934). De Acuña, who sailed the Amazon from Quito to Belém in 1639, called the Nhamundá River *Cunuris* or *Conduris*. He reported having encountered several native groups, such as the Apantos, Taguaus, and Cacarás, and also mentioned the Amazons.

The Konduri are often compared with the Tapajó, although they were said to have "their own idols, ceremonies, and government" (Heriarte 1874 [1662]:38). Heriarte mentioned that their subsistence was based more on manioc and game than that of the Tapajó. Fish seem to have been important to them, along with the use of turtles and manatees both for their meat and fat. During the colonial period, manatee fat was used to preserve various types of meat. Heriarte (1874 [1662]) also mentioned the growth of wild rice in their lakes, which they used to prepare wine. Konduri pottery was known for its high quality.

Before the Jesuits were expelled from the lower Amazon region in 1661, Father Manoel de Souza was sent to establish a mission among the Konduri (also referred to as Condurizes), along with another friar, 25 Portuguese soldiers, and 20 Amazonian natives. They were supposedly well accepted by the local people in the beginning, but the death of Father Souza is shrouded in mystery, indicating that he could have been killed by "the barbarians" (Bettendorf 1990).

The depopulation of the lower Nhamundá and Trombetas Rivers happened shortly after the first decades of more intensive contact. By 1691, when Father Fritz journeyed up the Amazon, Carvajal's province of Saint John had already vanished.

If information on the people who lived along the Amazon and the lower courses of the Trombetas and Nhamundá Rivers is sparse, there are no data at all on the populations that settled along the middle and upper courses of the same rivers.

Little by little, these places were occupied by populations fleeing from the Portuguese and the missionaries. In 1801, the Uaboi, who were living in a mission called "Village of the Nhamundá," reacted violently against the mission's rules and fled to the hinterland. During the nineteenth century, African slaves would also look for shelter along the Trombetas River, seeking the cover of its many waterfalls. Contact between runaway slaves and indigenous peoples was not always peaceful; there were frequent disputes over territories and also women. The *mocambos* were mainly populated by men, so they would kidnap native women for wives.

Although other groups are also mentioned, sources generally point to the Konduri as the first inhabitants of that region, with an occupation between 1541 and 1725, and direct contact only until 1691 (Guapindaia 2009). The knowledge that they produced high-quality pottery—which Bettendorf emphasized, although he did not describe the ceramics—is the only link between the ethnohistorical Konduri and the archaeological sites. In any event, after Barbosa Rodrigues (1875), scholars have assumed that the ethnohistorical Konduri occupied all the sites where ceramics of the Incised and Punctate Tradition are found, and the pottery thus became known as the Konduri ceramic complex.

## The Archaeological Landscape

The Tapajó and Konduri occupations as well as those by several other ethnic groups poorly described in the chronicles can be archaeologically identified in the form of large archaeological sites that feature, as their main landscape signature, the occupation of river bluffs, plateaus, and summit tops along the Nhamundá-Trombetas and the Tapajós Rivers, where abundant pottery remains are found throughout the deep black soil strata. There are also numerous smaller sites located along less important rivers and streams, especially between the Amazon River (Arapiuns River) and the left bank of the Tapajós River, as well as along the drainage system between the mouths of the Nhamundá and Trombetas Rivers. In Juruti, a city located on the right bank of the Amazon River, between the Nhamundá and Trombetas River mouths, large ADE sites have been found and studied in the scope of a salvage archaeological project (Costa 2009; Sousa 2008). Thirty-two kilometers east of Juruti, Nimuendajú located several small sites on the south margin of the Vila Franca Lake, which he interpreted as fishing stations (Nimuendajú 2004:131–135).

The configuration of these settlements on the landscape suggests the existence of interrelated regional systems. Surveys conducted between the courses of smaller rivers in Itaituba, situated on the left bank of the Tapajós River and some 200 kilometers south of Juruti, and the Nhamundá River located a route made up of rivers and canals, where 33 ADE sites containing ceramic sherds of the Incised and Punctate Tradition were found. These transversal routes might have been used for the flow of goods and people away from the main course of the Amazon River (Duarte-Filho 2010; Figure 4.1).

The two major sites (Santarém-Aldeia and Oriximiná) that were likely important political centers are located on opposite banks of the Amazon River. At the Tapajós River's mouth, the supposed Tapajó capital has been covered by the city of Santarém, whereas at the Trombetas River mouth, a large site lies underneath the city of Oriximiná. Few sites are placed next to major rivers like these. Along both the Tapajós and Trombetas Rivers there is an interesting system of lakes aligned to the main river courses. The lakes are connected to the rivers by a small canal, sometimes less than 5 meters wide. During the summer months, fish are abundant in these lakes.

In 1852, Bates (1892) reported a peculiar way to fish, common in the Tapajós River, but which he had not seen anywhere else: the first peoples' descendants used a liana called *timbó* (*Paullinia pinnata*), which would release a toxic gas when beaten against the water, impairing the fish's respiratory system. They used liana segments that were about 3 feet

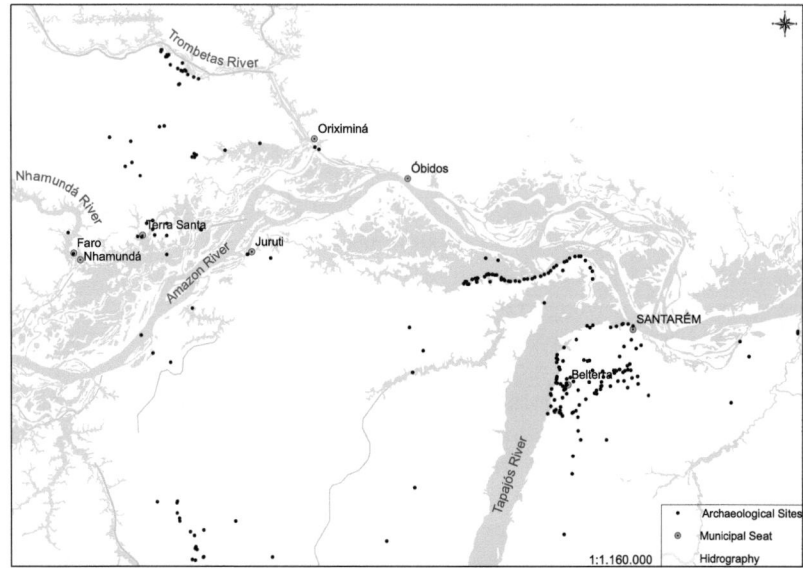

Figure 4.1. Archaeological sites containing Incised and Punctate Tradition ceramic styles.

long, and as the juice was released and stained the water, the fish would come up for air, floating on their sides with open gills, and thus be easily captured. This practice, as observed by Bates, was mainly used by women and boys. In 1768, Father Noronha (2006) reported the fishing of huge manatees in the Nhamundá lakes, especially to harvest their oil.

Aside from Vila Franca Lake, where Nimuendajú found many sites that could have been fishing stations, no other surveys have been conducted at the Tapajós lakes, in part because the area between the right bank of the Tapajós River and the plateau is a National Forest Reservation (FLONA Tapajós), created in 1974, and research there depends on special permission from the Brazilian Institute of the Environment (IBAMA). Even so, there is information on the existence of large ADE areas on the slopes that rise above the right bank of the Tapajós River, within the limits of the FLONA. Conversely, most of the archaeological research along the Nhamundá and Trombetas Rivers has been conducted around lakes, where archaeologists have found the largest ADE sites (Guapindaia 2009).

The first explorations of the Trombetas and Nhamundá Rivers date to the nineteenth century. Barbosa Rodrigues, a botanist, was the first to mention the ceramics found in cultivated plots, and to attribute them to the Konduri and Uaboi. He also mentioned stone figurines, which typically represent a man with an animal on his back (Porro 2010), for which the region became famous. Nimuendajú, during his fieldwork for the Ethnographic Museum of Gothenburg, was the first to actually record and map sites along the Tapajós and Trombetas Rivers. From April to August of 1923, he located 48 sites south of the city of Santarém, in the town of Alter do Chão and around Vila Franca Lake, which he believed to be less than half of the existing sites (Nimuendajú 2004). While traveling to the Nhamundá and Trombetas region, Nimuendajú found sites on a swampy area between the two rivers, where they are connected by a channel called Paraná do Bom Jardim. In this area, several small, flat mounds rising above the high-water mark were interpreted by Nimuendajú as dwelling sites. He visited four of those, finding ceramics and black soils in three of them.

He mentioned many other possible sites, but did not have time to visit them all. The largest site was Oriximiná, where he found ADE and a large amount of pottery sherds. These sites have not yet been studied. Nimuendajú reported finding the Sapucuá Lake, which resembles Vila Franca Lake. He found ADE sites on a hill located on the northern margin of the lake. Traveling along the Trombetas toward Óbidos, he found other sites at the Iripixy and Tapecurú lakes.

Although Nimundajú saw himself as an explorer and did not have a formal education in anthropology, his remarks on the similarities and differences between the Tapajó and the Konduri pottery, along with his observations on the probable limit between the two areas—which he placed in the Serra dos Parintins—are invaluable sources of information for current studies in the area.

Another important contribution to our knowledge about the Nhamundá-Trombetas River basins came from João Barbosa de Farias, an ethnologist who recorded sites between 1928 and 1929 and made accurate observations regarding pre-Columbian settlement patterns, stating that the sites were consistently located around lakes. He recorded 17 sites with ADE and pottery remains. He characterized the sites as having a layer of black soil (although he wrongly associated it with lake sedimentation processes). He thought the natives had a special connection with the lakes, which he believed were part of their religious cults. This was probably because of the ceramics found in those sites, displaying exquisite modeled figures thought to represent supernatural beings (Figure 4.2).

In 1952, Peter Paul Hilbert, an ethnologist from the Goeldi Museum, surveyed the lower Nhamundá and Trombetas Rivers, recording 41 sites. Hilbert stated that the native peoples who inhabited the region at that time did not recognize the vestiges found at the ADE sites as their own. These sites were located on river bluffs, lake margins, and summit tops. The sites were 1 to 3 hectares in area, with deposits 30 to 60 centimeters deep.

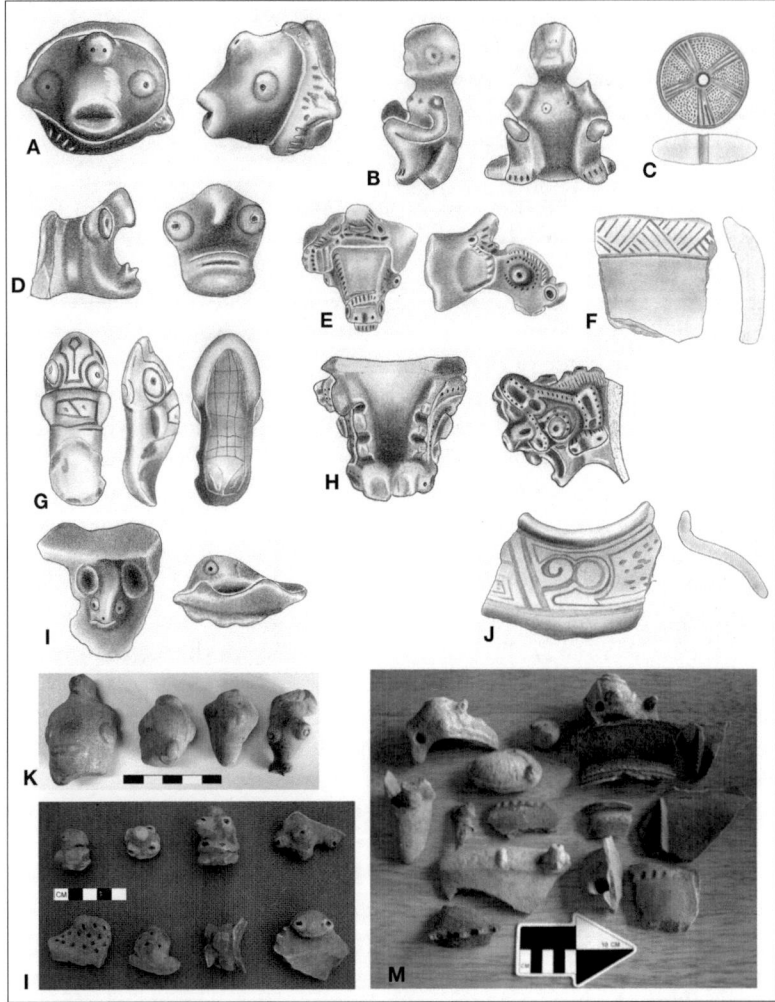

Figure 4.2. Incised and Punctate Tradition ceramics: (A–G) sherds from the municipality of Juruti sites; (H–I) sherds from the municipality of Parintins sites; (J, M) sherds from the municipality of Santarém sites; (K–L) sherds from the municipality of Belterra sites (drawing by Deise Lobo; photos by Marcio Amaral and Denise Schaan).

## The Tapajó Sites

In 2006 and 2008, we surveyed and mapped sites located south of the city of Santarém, especially along the Belterra Plateau, which have yielded important information concerning landscape modifications in that area, as well as a preliminary assessment of the pre-Columbian settlement patterns (Stenborg et al. n.d.). The Belterra Plateau is a Pliocene period formation composed of a material that is known as Belterra clay, which also occurs in other parts of the basin. South of the city of Santarém, the plateau rises between 130 and 180 meters above the water level, whereas on the opposite side of the Amazon River, north of the cities of Faro and Terra Santa and between the Nhamundá and Trombetas Rivers, it is around 60 meters high (Sombroek 1966). The geomorphologic similarity between these two areas had consequences for the human occupation of the landscape.

We have located so far a total of 104 sites on the Belterra Plateau and along the Tapajós River, mapping 44 of them with GPS. Twenty of these sites were visited and briefly described by Nimuendajú in the 1920s.

Almost all of the surveyed sites have been identified on the basis of their deposits of ADE soils containing ceramic sherds, most of which were used for growing soybeans, papaya, corn fields, and house gardens. Site sizes range from 0.1 hectares (Guari) to 16 hectares (Santarém-Aldeia), with ADE strata depth varying between 0.2 meters (Pindobal 1) and 2 meters (Vila Americana and Santarém-Aldeia).

The ceramics found at the sites, which were usually abundant on the surface, do not differ from the classic ceramics found at the Santarém-Aldeia or Porto sites. In the Belterra Plateau, there is a number of small sites separated by a narrow strip of unoccupied land (such as Lavras), which were probably several neighboring villages. The present dirt road was probably built on top of the old one. Besides ceramics and ADE soils, lithic artifacts are also abundant; the landowner of São Domingos also found a *muiraquitã* (frog-shaped pendant).

Some sites are located at strategic positions with ample views of the surroundings, such as the Curucurui and Capiranga sites (Figure 4.3C).

Aside from considerations on site location and size, we have proposed the existence of at least three types of sites in the area under investigation (Stenborg et al. n.d.):

(1) A few large sites situated at low altitudes, featuring high concentrations of archaeological materials, deep ADE deposits, and proximity to the main regional watercourses. These sites—exemplified by Santarém-Aldeia, Porto (Roosevelt 2000; Schaan 2010b), Lavras I, São Domingos II, São Domingos III, and Vila Americana—can be interpreted as large population centers.

(2) A few large sites containing much lower concentrations of cultural material and thinner ADE strata generally without many materials (in many cases of lighter color, referred to as *terra mulata*, after Kern, Ruivo, and Frazão 2009; Sombroek 1966); although they are situated near watercourses, their distribution is not limited to the vicinities of major waterways. These sites include Sitio Juá Jacú, Pindobal, Lavras II, Lavras III, Carapanari I, Carapanari IV, and Carapanari V.

Figure 4.3. (A) Excavation of a *bolsão* at the Port site, city of Santarém (photograph by Marcio Amaral); (B) Terra Preta do Jacú site, Belterra Plateau; (C) Serra do Capiranga site, in the back; (D) pond excavation at the Zinha site, Santarém municipality (photographs by Denise Schaan).

Their period of occupation was considerably shorter than that of the first category of settlements, so they might be later sites, belonging to a period of expansion of the regional system.

(3) Small sites, with ADE deposits of varying depths, a high concentration of archaeological artifacts, and often located on hills, with limited access to surface water. Sites such as Genipapo I, Genipapo II, Bom Futuro, and Posto Novo can be placed in this category. These sites probably belong to different time periods, and their functions may have changed over time.

In some of the sites situated on the plateau, Nimuendajú (2004:Plate 203) located (and we confirmed) the existence of artificial wells, which were vital in a landscape devoid of watercourses. The distance between these settlements and the Tapajós River is about 8 kilometers. The wells range from 15 to 30 meters in diameter, but in one of the sites we found an excavation that exceeded 100 meters. When Nimuendajú located the wells, they were still in use by the local population. However, during our survey in 2006, we heard from many people that pesticides used in agriculture had contaminated the water in the well, and it was not safe to drink anymore. Nimuendajú also called attention to roads that connected one black soil area to the other, which we were not able to locate, given the fact that those roads were likely reworked into present ones.

So far, survey results indicate that neither ADE sites nor archaeological sites in general are more abundant in the vicinity of the Amazon River than along the lower Tapajós and in inland areas. Rather, there are clusters of sites in the southwestern part of the surveyed area, near the city of Belterra, as well as in some inland areas, such as the one formed by the Lavras 1–3 and Jacú sites.

None of the sites mentioned here were excavated. The most reliable information on archaeological deposits related to the Santarém culture comes from excavations at the Port site, in the city of Santarém.

The Port site is a large 35.7-hectare ADE area. Archaeological deposits are 2 meters deep in some parts of the site, but not all of them are related to the Santarém occupation. In deeper deposits, Roosevelt has identified a Formative period stratum, and possibly a pre-Santarém society not yet studied (Schaan 2010b). Although Roosevelt has been studying the Port site since 1987, little has been published since then. She has chosen Port instead of the Aldeia site because the latter lies underneath modern buildings and paved roads. Although construction and soil removal at the Port site have affected the archaeological deposits, the existence of large open areas has allowed for controlled excavations.

The information obtained through excavations has been compared with a series of observations made by Amaral, an amateur archaeologist who has participated in the lower Amazon Project under Roosevelt since 1987, and has recently joined our staff as a technician. In the mid-1990s, the construction of a terminal at the Port site involved excavations that revealed profiles observed and registered by Amaral, who also collected material culture that was being thrown away. These profiles contained floor structures that were aligned in the north-south direction with widths that varied between 8 and 12 meters, and forming strata that were up to 2 meters deep. It was not possible, however, to determine the extension of the habitations, which were on artificial platforms. The hard dirt floors were renovated periodically, causing a gradual elevation of the house floors. Amaral has observed contemporary houses that use similar technology for house floors. While excavating an archaeological site last year on a Tapajós River bluff in the Aveiros municipality, we found a house that has sheltered a family of 12 for 30 years. The hard dirt floor has been constantly mopped and has been rebuilt occasionally, as the need arises; this practice has caused an elevation of approximately 40 centimeters throughout the years, which can be observed in an area of the house profile where the wooden walls rotted (A. M. A. Lima 2010).

In other parts of the site, where stack floors are not present, we have found features such as *bolsões*, a possible consequence of the termination rituals that took place in open plazas. While looking for ceramics in the backyard of several houses in the city of Santarém (Aldeia site), Barata discovered that most of the vessels were deposited in such cavities (1953). Ritual *bolsões*, although little-studied, are structures that concentrate large amounts of a variety of cultural and organic materials, from which broken vessels can sometimes be reconstructed. They are round, vary between 2.5 and 4 square meters in size, and extend to a depth of 70 to 100 centimeters (Figure 4.3A). The color of the soil inside the *bolsão* is usually a lot darker than that of the outside, indicating the presence of organic debris and charcoal. The ceramic assemblage found inside includes red-slipped, incised, zoomorphic appendages; broken pieces of figurines; fragments of ordinary plates and bowls; bone fragments; charcoal; lithic artifacts; rocks; and burned clay. In one of the *bolsões*, we have found a broken clay stove (Schaan 2010b).

Our 2009–2010 excavations at the Port site have also yielded two funerary vessels that were collected in a block of hard sediment, and which are being currently analyzed by a physical anthropologist. We had previously observed rims of buried broken vessels at the Port site in an area that had been bulldozed to build a soccer field. This was highly unexpected, since all the ethnohistorical data indicate that the Tapajó practiced cremation and the mummification of bodies, and there had previously been no record of urn burials.

## The Konduri Sites

Settlement patterns in the Nhamundá-Trombetas River basins have to be inferred from the works of Nimuendajú (2004), Hilbert and Hilbert (1980), and Guapindaia (2009).

In the late 1970s, Peter and Klaus Hilbert (1980) were conducting a survey in the Trombetas-Nhamundá region when they identified 11 sites, all characterized by ADE deposits, which allowed them to define the local cultural sequence. The sites were located along the margins of rivers and lakes bluffs, reaching 40 to 50 meters in height from the high tide level to the summit tops. The depth of the archaeological deposits ranged from 30 to 60 centimeters. They also reported finding a site on the plateaus that was located 175 to 180 meters above sea level. During the excavations, they found an early ceramic complex that preceded the Konduri, which they called Pocó. Pocó sites date between 65 BC and AD 205. The type site of Pocó is located on the right margin of the Pocó River, an affluent of the lower Nhamundá River. Pocó ceramics resemble the Barrancoid Tradition found at the Orinoco River mouth: the ceramic paste is mixed with organic temper (*caraipé* and *cauixi*), and vessels are carinated and painted red with modeled adornos.

The relationship between Pocó and Konduri sites has not been well established. The few available dates indicate a gap of a thousand years between them, whereas the ceramics found in archaeological deposits appears to be partially contemporary. This might be a matter of misinterpretation of the depositional processes. Although both ceramic complexes seem to relate to the same old tradition, they are clearly distinct.

Konduri sites in the area are found in the upper deposits, sometimes over the Pocó occupation. Peter and Klaus Hilbert conducted a deeper investigation at the Boa Vista site, located on a high bluff at the right margin of the Trombetas River, some 80 kilometers from the Pocó site. They reported finding deep and shallow bowls as well as open-rimmed vessels in the Konduri deposits; they also found manioc graters and vessels with tripod, pedestal, and annular bases. The typical conical supports were 3 to 15 centimeters long, sometimes with biomorphic decorations. There were arc-shaped handles over the vessels, with incised straight lines for decoration. They also emphasized the abundance of punctates and fillets, also finding red and white slip as well as red paint on white slip, along with other less common types of decoration. The lithic assemblage comprised basalt axes and large ferrous sandstone grinders.

Archaeological research has placed the Konduri occupation between the thirteenth and the seventeenth centuries (Guapindaia 2009). These societies occupied riverine and lakeshore locations (where high bluffs were preferred), but also the interfluves, where sites are smaller, probably representing seasonal campsites. Based on her own as well as previous research, Guapindaia counted 120 sites in the Nhamundá-Trombetas region. In the 1980s, a mining company was established in the vicinity of the Trombetas River; to protect part of the forest, the government created the Saracá-Taquera National Forest. A cultural resources management (CRM) project has been running since then, led by archaeologists affiliated with the Goeldi Museum from the city of Belém. Seventy-eight sites have been recorded just in the National Forest and within a 10-kilometer radius, of which 38 are located on lake shores or within 2 kilometers of them; 26 are situated on the lowlands between the Trombetas and other main rivers; nine are on the main riverbanks, notably on bluff tops; and five are on plateau tops (Guapindaia 2009:86). The most important landscape features are the lakes located along the Trombetas River, which are incorporated into the river during the high tide season due to the overflow of the small canals that connect them.

Guapindaia notes that the distance between lowland sites located at the interfluves and those situated on lake and river shores ranges between 30 and 50 kilometers, which might not have impaired their ability to remain in contact; this finding agrees with ethnohistorical accounts describing the relationship between riverine and hinterland peoples, raising the possibility that they might have been part of a same political economy.

The location of sites in the plateaus was not determined through systematic surveys, but in accordance to the needs of the project. Additional sites probably in the plateaus were not surveyed. In any case, out of the 12 plateaus surveyed, only two had sites; these sites are smaller in size, have shallower cultural deposits, and in some cases do not feature the typical ADE soil, thus likely representing campsites used for the collection of fruits and nuts.

According to Guapindaia (2009:86), the eight sites located at river margins have the largest areas, between 1 and 4 hectares. Lake margin site sizes have a wider range of variation, between 0.28 to 1,500 hectares. The largest sites here might actually be several adjacent and overlapping sites.

Guapindaia estimated that the plateau sites have areas ranging between 0.4 and 15 hectares, whereas the areas of sites located at interfluves vary between 0.11 and 27 hectares.

The geographical distribution of the sites indicates that relatively dense populations lived in areas rich in aquatic resources, whereas the density of ADE deposits points to settlement patterns that are similar to those in the Tapajós River region. In both places, there are still insufficient data to understand the distribution and functions of the ADE sites and adjacent *terra mulata* areas (supposedly agricultural fields) because neither chemical nor morphological studies of the soils are available. The distribution of sites in the landscape suggests a settlement pattern related to the exploitation of aquatic resources, although further studies are needed to establish site function and the possible relationship between sites.

Unfortunately, archaeological research conducted at Trombetas River sites has been historically concerned with collecting ceramics for seriation and other types of analyses, so we know little about cultural features. Moreover, excavation in ADE sites requires that close attention be paid to discrete layers, observing differential deposition, soil texture, and the collection of microremains, since soil color can be the same in several layers, making them indistinct to untrained eyes.

## Comparing Sites along the Tapajós and Nhamundá-Trombetas River Basins

Comparing sites located at the Trombetas and Tapajós River regions, it is possible to conclude that the similar distribution of sites in the landscape, along with the presence of Incised and Punctate Tradition ceramics, indicates cultural and likely political proximity between the two populations. Sites located south of Vila Franca Lake as well as in Itaituba also feature ceramics related to the Incised and Punctate Tradition that belong to styles similar to both. Current research in these two areas as well as in the Santarém and Belterra Plateaus will enable us in the next few years to refine our hypothesis of a large settlement system encompassing this large region.

The number and size of ADE sites located in the Tapajós and Trombetas regions indicate a fairly high population; those were probably important polities by the fifteenth century. At these sites, in addition to the elaborate pottery, evidence points to the production of stone tools and stone adornments that were traded with other Amazonian polities, which put the Tapajó and the Konduri as the critical node in the long-distance trade network system in the basin.

## Amazonian Dark Earths

The Tapajós and Trombetas River settlements are characterized by an abundance of ADE deposits that can be as deep as 2 meters, as is the case at the Santarém-Aldeia and Port sites. For this reason, it is important to discuss the processes that formed these sites, their particular stratigraphy, the intentionality or not of their construction, and current preservation issues.

ADE sites were initially called *terra preta de índio* (Indian black soil) or *terra preta arqueológica* (archaeological black soil) by scholars, given their obvious association with indigenous settlements (Kern 1988, 1996; Smith 1980; Sombroek 1966). They soon came to be a signature for archaeological sites in the Amazon basin, since black soils would consistently be found in association with ceramic sherds. At first, some thought the ceramics were there because the indigenous peoples would have preferred those areas for habitation, as the soil was extremely fertile (Myers et al. 2003). This assumption stemmed from the fact that these sites were being chosen by people in the twentieth century for planting varied crops, given their high yield. However, soon it became clear that those soils did not occur naturally, but were formed as pre-Colombian populations lived on them, discarding organic matter: in other words, they were "kitchen middens" (Sombroek 1966).

ADE has a different chemical signature which distinguishes it from other Amazonian soils. It is composed of higher concentrations of chemical elements such as phosphorus, calcium, magnesium, zinc, and manganese compared with original, adjacent soils, and feature a higher pH as well as higher amounts of organic material, which are the reasons for their high fertility in agricultural cultivation (Kern and Kampf 1989; Kern et al. 1999; McCann et al. 2000). High levels of phosphorus, calcium, and magnesium in ADE soils can be attributed to animal residues such as bones, blood, carapaces, and feces, whereas zinc and manganese are probably from the decomposition of palm trees, the leaves of which were commonly used as a construction material in house walls and roofs, beds, hammocks, and basketry in general (Kern et al. 1999). In an upper Xingu River village, Schmidt and Heckenberger (2009) have found high levels of sodium and iron in soils where manioc processing was the main activity.

The vegetation that grows on ADE soils is very distinctive and easily identified. In ADE soils, it is possible to find several plant species currently used by human populations, such as cacao (*Theobroma cacao*), papaya (*Carica papaya*), cupuaçu (*Theobroma grandiflorum*), Brazil nuts (*Bertholletia excelsa*), and kapok (*Ceiba pentandra*; Woods et al. 2000). Most of these species are found in abandoned indigenous plots along with palms and bamboos, and are thus involved in what William Balée (1989) has called cultural forests, or anthropogenic forests.

Most ADE soils, however, do not indicate old agricultural fields, but rather pre-Columbian indigenous villages. This is because ADE sites have plenty of cultural features and artifact remains such as floors, stoves, garbage dumps, postholes, and burials, as well as an abundance of ceramic, lithic, bone, botanical, and charcoal remains. Since cultivation fields are not likely to display these types of remains, it is clear that these places were originally used for habitation, feasts, and ceremonies. Moreover, at most sites it is possible to distinguish differences in the extent and depth of ADE soils, which is a result of the range and chronology of cultural activities. For instance, deposits of ADE are rare in places where a central plaza is constantly swept of cultural debris. Conversely, backyards where garbage is usually discarded will contain large and deep deposits of ADE.

Although ADE provides a good signature for cultural activity areas within a site, a lot of noise will be produced in the archaeological record if there are multiple episodes of reoccupation or changes in the village layout through time; in these cases, it might be difficult to determine which deposits are contemporary and which were produced by different societies (Erickson 2003). So, although ADE is an excellent correlate for population density, there is no direct link between these two variables. For example, the population density in a 3-hectare Marajoara mound could be that of 200 people over the course of 500 years, but the ADE deposits in a typical ceremonial mound are negligible, since garbage would be consistently thrown off the mound, keeping the habitation areas and plazas clear of debris (Schaan et al. 2009a).

In the periphery of and between ADE sites, large areas with soils that are lighter but still more fertile and darker than adjacent soils could be due to the intentional manipulation of areas for cultivation (Woods and McCann 1999). Accordingly, these soils would have been subjected to the addition of mulch and charcoal or ashes, lowering the pH and increasing fertility. Sombroek (1966) called those areas *terra mulata* (brown earth), believing that they resulted from their long use as cultivation plots.

Although scholars agree that ADE soils were produced by indigenous societies, there is no consensus in regards to either their intentionality or origin. Another important point of debate is the process of ADE formation. Although it is clear that organic matters and charcoal were incorporated into the soil, it is still a matter of debate why these soils became so resistant to leaching, and why their chemical components remain present and the soil is fertile even after years of continuous use. One of the possible answers to this question is the presence of black carbon (BC), introduced to the soil through controlled burning. BC is highly resistant to microbial degradation, and thus maintains its properties over millennia (Schmidt and Noack 2000). In soil, BC would be responsible for maintaining high pH levels, reducing leaching by increasing cation exchange capacity, and stabilizing chemical elements that improve soil fertility (Glaser et al. 2002). In most Amazonian soils, the continuous use of soils for agriculture led to a rapid loss of organic matter and fertility, whereas in ADE soils, fertility is found at greater depths, which may be credited to BC presence (Tsai et al. 2009).

Woods and McCann surveyed several ADE sites in the Tapajós region, arguing that many of the sites they found did not conform to the midden model (daily accumulation of domestic refuse), and contained soils that would have been prepared by humans for cultivation. They believe that the dark color of the soils was an "indirect result of the chemical changes that stimulated soil biota activity and 'growth' downward through the incorporation of their organic byproducts" (Woods and McCann 1999:2). In this sense, the depth of the black soil would not result from time of occupation, but from other variables such as "geomorphic context and soil texture" (Woods and McCann 1999:2).

With so many variables influencing soil composition and behavior through time, we need a better control on chronology and a closer look at the associations between artifact, feature content, and chemical signatures. In fact, we need more research involving extensive excavations as well as the interpretation of features and soil analysis. Ethnoarchaeological research as well as the experimental production of ADE have much to contribute to solve archaeological problems.

Recent studies conducted at ADE sites have sought to better define the tempo of ADE formation and its relation to crop domestication, sedentary life, and population increase. Manuel Arroyo-Kalin (2010) has studied the formation of ADE in the central Amazon, where he found evidence that soil enrichment started during the mid-Holocene period, before most ADE sites were formed. The presence of charred remains and chemical signatures in pre-ADE soils, contemporary with manioc domestication, would substantiate the relationship between settlement nucleation, agriculture, and the formation of ADE soils, a phenomenon that would later lead to the formation of large ADE sites related to complex sociopolitical systems. Although it is clear to me that manioc has been the main staple for Amazonian populations since its inception as a domesticated crop in the basin, I do not believe ADE soils were formed primarily as a result of expansion and intensification of agriculture.

Since most were formed during the period of population increase and the formation of regional societies beginning in the first millennium, I believe ADE are linked to a process of landscape transformation, which includes a change in the subsistence systems as well as an intensification in the production of material culture related to ancestor cults. This change in the subsistence systems would have involved the massive capture of fish in fishponds, fishweirs, and dammed/poisoned lakes, with the transformation of this resource into dry fish and fishmeal for storage. Given the high phosphorus content found in fish, this production process would have contributed to the addition of fish remains to the soil (Erickson 2003; Lehmann et al. 2003; Sombroek et al. 2003). At the same time, the increased production of ceramics in the villages required huge amounts of fuel for the firing process, which would be obtained by burning wood, which in turn would ultimately be added to the soil as charcoal.[1]

## *Muiraquitãs* and the Greenstone Trade

The term "greenstone" is generally used in a broad sense, referring to diverse materials that include jade (and its variants jadeite and nephrite), serpentine, agate, chalcedony, opal, amazonite, quartz, and basalt. Mineralogical analyses performed on greenstone pendants from the lower Amazon River have shown that most were made of the nephrite variety of jade as well as from other greenstones that simulate jade, but that might be of local origin.

Some of these minerals are more rare than others, but the green color seems to have carried cultural and social meanings shared by many societies throughout the pre-Columbian Americas. Greenstone is actually the popular name of an igneous basalt that contains green minerals such as chlorite and epidote (Atwood 1997). This green variation of the common basalt, which was more easily obtained than the valuable jade, may have been used in the production of the desirable objects as a substitute for other finer rocks. The green rocks used by Amazonian peoples actually come in a variety of colors, from light green to brownish and dark green, and sometimes almost white or black, depending on their mineral components (Guerrero 1998). These rocks were used to produce tools, figurines, pendants, and other ornaments.

The term "jade" was first used by the Spanish when they came in contact with the Aztecs' greenstones, and was later applied to the oriental jade, which arrived in Europe in the seventeenth century (Easby 1968). The Spanish word is derived from the term *piedra de illada*, or loin stone, a reference to its curative properties in treating loin and kidney diseases (Balser 1974; Boomert 1987; Easby 1968; L. Lima 2010). The Aztec used the Nahuatl term *chalchihuitl* to refer to jade (Boomert 1987; Easby 1968; Tibón 1983). *Chalchihuitl*, or the water goddess, was supposed to be the same color as water and vegetation (Tibón 1983). The curative properties attributed to jade ornaments are common everywhere they occur (Boomert 1987; Easby 1968), which speaks of a shared and basic cultural meaning that might have had local variations.

Ethnohistoric sources report that the lower Amazon River was famous for its greenstones, and that the Tapajó manufactured and traded fine pottery, hammocks, drinking cups, wooden seats, *muiraquitãs*, beads, and other figurines (Boomert 1987). *Muiraquitã* is an indigenous word that makes reference to its common frog shape (Barbosa Rodrigues 1899). The term *muiraquitã* seems to be related to frog-shaped pendants, but occasionally

Figure 4.4. Stone *muiraquitãs* from the Museu do Forte collection (photographs by João Ramid). Bottom right: a ceramic *muiraquitã* from Piquiatuba site, Tapajós River margins (photograph by Marcio Amaral).

people refer to any green pendant as *muiraquitã*, some representing birds, snakes, lizards, fish, turtle, and other animals. They may be cylindrical, rectangular, or bar-shaped (Scatamacchia 2000; Figure 4.4).

The archaeological record shows that *muiraquitãs* were more numerous in the city of Santarém, where the Tapajó society, with its strategic position at the Tapajós River's mouth, may have controlled part of the trade networks along the Amazon. There is no information on the source of the pendants. The fact that large quantities were found in the Tapajós River basin suggests local origin according to distribution models of trade goods (Shennan 1982).

Similar objects have been found in other areas, such as the Trombetas River and on Marajó Island, where fewer than five pieces have been reported, one representing a bat, two as anthropomorphic figurines, and two shaped like frogs (Banco Santos Collection, Museu de Arqueologia e Etnologia, São Paulo University).

The Tapajó actively participated in trade networks, and they likely had a prominent role in the trade of basalt axes, since there is evidence of a source of raw materials and a workshop south of their zone of influence (Gomes 1997). In sites located some 200 kilometers south of the city of Santarém, we have found fragments of stone axes in various stages of the production process (Silva In press), which suggests that they were one of the sources of manufactured products for exchange.

In the lower Amazon River, *muiraquitãs*, along with basalt axes, were likely the most important items in ceremonial exchanges among elites (Gomes 2001; Scatamacchia 2000). These objects, which were perforated so they could be attached to a supporting string, were probably worn by elite women (Gomes 2001), but several lines of evidence suggest their use as currency (Boomert 1987). Other perishable items, such as feathers, exotic animals, cloth, and ceramics, were probably highly valued and traded, but there is little information about them.

Workshops related to the production of such items have been found in sites along the Trombetas River; special tools for perforating pendants have also been found at the Santarém-Aldeia site. Greenstone adornments were commonly traded among the Amazonian elites, and were possibly used in the exchange of wives, as a bridal gift. The distribution of these prestigious items, which covers a broad area from the Caribbean to the Amazon basin, implies a connection between complex societies separated by great distances (Boomert 1987).

At the end of the nineteenth century, Barbosa Rodigues reportedly heard from an old woman in the city of Santarém that when she was little, the Tapuyos (Tapajós) traveled to the Nhamundá River annually to exchange goods for *muiraquitãs*, which had religious significance to them.

Ethnohistoric sources point to the region between the Trombetas and Tapajós Rivers as the origin of the greenstone pendants, which reflects the distribution of the archaeological objects, but not necessarily the source of raw materials (Scatamacchia 2000). Scholars have reported the occurrence of nephrite in the states of Bahia, Rio de Janeiro, São Paulo, and Paraná (Costa et al. 2002). Geologically speaking, however, the presence of nephrite in the superior courses of the main Amazon tributaries is a possibility (Costa et al. 2002).

Moreover, it is known that the natives themselves told the Europeans that the greenstones were obtained through trade with the tribe of the "women-without-men" (the mythical Amazons), located farther inland. According to the myth, the *muiraquitãs* were made of the green mud found at the bottom of a lake. The Amazons would model the objects underwater, and after completion would take them out of the water, where they would harden out of contact with the air (Heriarte 1874 [1662]). Since they did not have men in their tribe, the Amazons met men from other tribes once a year to reproduce. In those occasions, they would trade the greenstones for a variety of objects (Boomert 1987; Carvajal 1934).

In addition to the small pendants, small- or medium-sized figurines representing a human figure with an animal on its back also have their origin around the Trombetas/Nhamundá and Tapajós Rivers (Figure 4.5). Fewer than 30 of these statues have been found,

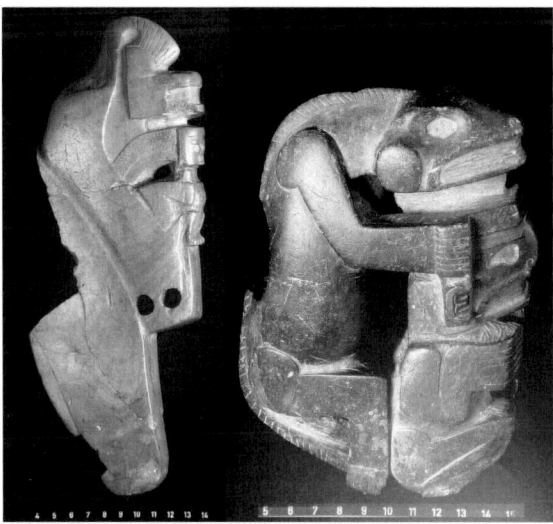

Figure 4.5.  Stone idols, Óbidos (photograph by Ferenc Schwetz; courtesy of the Museum of World Cultures, Sweden).

and today they are distributed among several different museums and private collections. These figurines have been called "stone idols," and they vary quite a bit in terms of size (20 to 70 centimeters) and iconography. Their only common feature is the existence of two holes, which were likely used for affixing to a supporting material, perhaps the bow of a boat. The function of these stone idols would have been symbolic, like a good luck charm in fishing expeditions or wars, for example, as a passage by de Acuña (1859) might suggest: "When they are going to war, they carry an idol in the bow of their canoes, in which they place their hopes for victory; and when they go out fishing, they take the idol, which is charged with dominion over the waters" (de Acuña 1859:83). Since de Acuña does not describe the idols or their raw material, we cannot know for sure whether he was referring to the stone figurines or not.

Recently, Porro (2010) has proposed a novel approach to these objects. He compared them with ethnographic objects related to shamanic practices with same iconographic motives, which associated a human figure and an animal, mirroring the transformation of the shaman in a powerful animal, such as the jaguar. Amazonian trumpets that predate the twentieth century would exhibit the same motifs. The most appealing argument in Porro's hypothesis is the use of the stone idols' holes for supporting inhalation tubes, which finally provides a more credible explanation for such features.

The iconography of these objects does not follow the local style. The ones representing people and beasts bear a greater resemblance to objects found in Nicaragua and northern Colombia, as some authors have suggested. The paradox, however, is that the idols have not been found outside the Nhamundá-Trombetas/Santarém area. Would they have been the product of trade? A mineralogical study of the idols as well as a more in-depth iconographic analysis would be necessary to precisely connect them to any local or foreign style.

## Tapajó and Konduri Ceramics

Similar to other archaeological cultures in the Amazon region, ceramics constitute the bulk of artifact assemblages at the sites of Tapajó and Konduri cultures, with the less-numerous, albeit important, remains of lithic tools and other lithic industries.

Archaeologists rarely recover whole vessels and objects, finding mostly potsherds. As for Tapajó culture ceramics (also called Santarém phase), iconographic studies as well as those dedicated to describing the ceramic industry have relied upon the examination of museum collections, from the twentieth century through acquisitions from private collectors in the city of Santarém. Many complete vessels were recovered from the city streets, where in 1922 torrential rains unearthed pottery and stone tools. The locals were accustomed to finding pottery in their backyards, which they called *caretas* (masks) or *panelas de índio* (Indian pots). Further excavations at disturbed sites around modern houses yielded more ceramics for museum collections (Barata 1953).

Whole Konduri vessels are even more rare given the scarcity of fieldwork and the fact that sites have not been looted. Funerary vessels have not been found at Konduri sites, and have rarely been found at Tapajó culture sites. That might be the result of sample bias and insufficient research.

Scholars who have dedicated themselves to the study of the Tapajó culture—also called Santarém phase—and Konduri phase ceramics have proposed various typologies, seriations, and interpretations. Rather than compare or revise them here, I will briefly describe the ceramic styles and discuss the use of pottery vessels in ceremonies as well as their possible sociopolitical and gender meanings.

Both ceramic styles were classified as belonging to the Incised and Punctate Tradition, which, based on similarities to Venezuelan styles, was initially dated after AD 1000. There are a few carbon 14 and thermoluminescence (TL) dates associated with Santarém and Konduri pottery that place both complexes within the AD 1000 to AD 1600 period (Gomes 2001; Guapindaia 2009; Hilbert and Hilbert 1979; Pouguet 2002; Quinn 2004).

Potters from both complexes used a fluvial sponge called *cauixi* (*Parmula batesii*) as temper material, and eventually a combination of *cauixi* and grog (crushed potsherds). Santarém phase vessels are more resistant than Konduri phase ones because the latter have higher amounts of *cauixi* (Guapindaia 1993). We have also found the use of *caraipé* (siliceous bark ashes) used as temper in combination with *cauixi* in sites located in the vicinities of both domains.

In both complexes, vessel surfaces are a pale brown color, and fragments feature a grayish core, with hue variations from pale brown to strong gray. The diagnostic feature is the profuse use of anthropomorphic and zoomorphic adornos (sometimes with double heads), generally as handles or simply as decoration. The animals more commonly represented are king vultures, caimans, agoutis, monkeys, frogs, bats, common foxes, birds, and jaguars.

In the Santarém phase style, the use of punctated designs is moderate, whereas in the Konduri phase pottery they are abundant, along with nubbins and modeled forms which sometimes cannot be morphologically compared to the representation of particular animals or humans. Decorations in Konduri phase ceramics are coarse, whereas in the Santarém phase style they seem more carefully produced. Both make use of incisions, usually as straight lines applied in repetitive patterns inside bands that surround the vessel or plate. Curved lines are used only in Santarém phase ceramics. The fish spine motive is found at Konduri phase sites. Both styles have double rims with no apparent utility, but hollow rims are only present in Santarém phase ceramics. Konduri phase vessels have tripod bases, which are absent in Santarém phase assemblages, where annular bases are more popular. The use of paint (mostly red, but also red and white slip) is present in both complexes, but Santarém phase painting is more resistant to water. Basketry motifs appear in both pottery styles, probably deriving from a similar industry that used mats as supports for vessel shaping. Doughnut-shaped eyes are also a common characteristic in animal and human representations.

Hilbert identified three different ceramic styles at Konduri phase sites, which he called Konduri, globular, and sand-tempered. The globular style is included in the Barrancoid Tradition, occurring at the Sapucuá Lake sites and at the Oriximiná, Terra Santa, and Faro sites. He describes the vessels as hard, with great amounts of organic temper, the signature trait being a globular modeled adorno. The Konduri style occurs along the Trombetas River (for example at the Boa Vista site), and is characterized by the profuse use of punctated decorations with punctated fillets, zoomorphic adornos, pitted bulges, and incisions. Sand-tempered ceramics are coarsely finished, decorated with nubbins or anthropomorphic adornos, and sometimes with fish spine incisions; they are found at the Trombetas River mouth and at the Sapucuá Lake sites.

Although several scholars have studied Santarém phase iconography, none have dedicated themselves to the examination of Konduri phase iconography, probably because museum collections, which are the basis of such studies, are nearly devoid of Konduri specimens.

## Santarém Phase Iconography

Santarém phase assemblages have a variety of vessels decorated with incisions, occasional red and black paint over white slip, and modeled anthropomorphic and zoomorphic handles and appendages. Among these, the caryatid vessels, *gargalo* (neck) vessels, effigy vessels, and anthropomorphic figurines (Figure 4.6) have been the focus of studies.

Caryatid vessels (after Barata 1950) are three-part vessels that comprise a top straight-walled, concave-based small bowl supported by the heads of three female figurines in a squatting position, placed in equidistant positions, and whose feet rest over a pedestal base. This base can be described as two bowls joined by the bottom. Both the upper and lower parts of the caryatid vessels are ornamented with incised designs. The three figurines are using their hands either to cover their eyes, ears, or mouth. The position of the figurines varies from vessel to vessel, but within the same vessel they are always identical. Other anthropo-/zoomorphic adornos (usually four or five pairs) are applied over the top

Ponds, Lakes, and Feasts: The Cultural Geography of Anthropogenic Soils   133

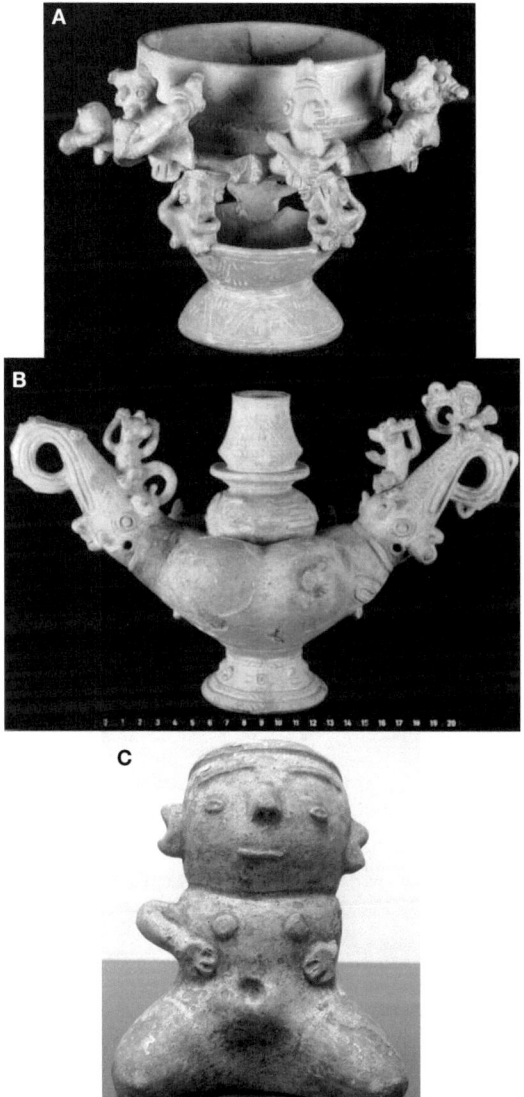

Figure 4.6. Tapajó pottery: (A) caryatid vessel; (B) neck vessel (photograph by Ferenc Schwetz; courtesy of the Museum of World Cultures); (C) female figurine 12 centimeters in height (Nimuendajú Laboratory collection, Santarém; photograph by Marcio Amaral).

bowl. These figurines are either single- or double-headed. The latter type features a human and an animal head. The single-headed figurines are double-faced, representing a human or an animal depending on the observer's point of view. There are only a few complete specimens of this type of vessel in museum collections, but broken figurines are present in most ceramic assemblages found at Santarém phase sites.

The *gargalo* vessels comprise three main parts: an annular base; a complex globular body with elongated handles, profusely ornamented with modeled zoomorphic adornos; and a superior neck. Barata (1950) defined two types of *gargalo* vessels. The first has a more exotic body generally made up of four concave parts, from which two elongated handles emerge from opposite sides of the vessel, formed by complex caiman or bird heads. The four feet of two realistic three-dimensional frogs are also placed on opposite sides of the vessels, in between the caiman adornos. Several other human or animal heads as well as stylized snakes can be found in between the caimans and the frogs. The vessel's neck is placed as a hat over an anthropomorphic head, sometimes with double flaps.

In the second vessel type, the neck/hat and the annular base remain, but the vessel's body assumes the realistically represented shape of an entire animal (turtle, frog, jaguar) or a human. On occasion, feet or paws replace the base. I believe that this vessel type represents a mythical tale. Barata (1953) noticed that incisions applied over the vessels are not devoid of meaning, showing that they are stylized representations of frogs, snakes, and owls.

Figurines (mostly broken) are commonly found in ADE, especially in *bolsões*. All are hand-modeled (they did not use molds), and some are hollow; pebbles were commonly placed inside the hollow figurines so that they would make noise when shaken. Their size varies between 3.2 and 28 centimeters. In Guapindaia's study (1993), 97 percent of the figurines were anthropomorphic (34 percent represented females), and 52 percent had severed heads. which may mean these objects were ritually broken. Unlike the Marajó figurines, however, Santarém phase figurines are more realistic, in some cases featuring the person in a standing position.

Female figurines can be identified by the representation of the genitals (an incised triangle or rectangle, a vertical incision, or a *tanga*), breasts, and sometimes pregnant bellies. Even male figurines might have breasts (Guapindaia 1993). The body is ornamented with incised lines, which likely mimic body painting or scarification (Barata 1950). An intriguing type of figurine represents a human head on top of a single leg-foot pedestal, which is called a unipod. Figurines sometimes have perforations on the top of their heads, where feathers were probably inserted; the same is believed to be true for some vessels (Barata 1950).

Pipes also appear in Santarém phase ceramic assemblages, but are absent in Konduri phase ceramics. Museums have large collections of Santarém phase pipes, which seem to be of indigenous production (Barata 1951; Linné 1928), although some decorative styles (leaves and volutes) suggest some production during the Colonial period. Moreover, their shape is very similar to that of Western pipes. Barata (1951) classified them into four different types (1) plants; (2) less skillfully designed plants; (3) Tapajós/anthropomorphic and zoomorphic; and (4) simpler plants), all contemporary to European presence among the natives.

## Gender in the Santarém Phase Iconography

Humans in Santarém phase ceramics are mostly represented in figurines and effigy vessels. Anthropomorphic figurines (mostly females, although males are also represented) are abundant in museum assemblages and in excavations. Effigy vessels represent realistic humans either engaged in some activity or placed in particular positions. Both types of representations generally feature naked people, and can be either males or females with body painting and adornments such as earplugs, headdresses, bracelets, and collars. The presence of *muiraquitãs* (greenstone amulets) on female headdresses suggests that those vessels portrayed elite women (Gomes 2001). In some specimens, the female sexual organ is represented with naturalism. However, female figurines are generally smaller and more stylized than effigy vessels, which more often portray males in a naturalistic fashion (Gomes 2001). This differential representation of males and females in Santarém phase pottery instigates an analysis of gender relations.

Gender complementarity and gender equality are commonly interpreted as characterizing mid-range, less-complex societies. Comparative studies, however, which examine the range of cross-cultural variability of gender roles in kin-based societies, can demonstrate that differences are partly due to particular cultural matrixes and low population levels (Crown 2000). Accordingly, small-scale societies, even when hierarchical, may not rely on gender as an important principle of social differentiation (Oyewùmí 1997). However, during processes of population growth and increased specialization, gender may become important in defining roles in contexts such as economic production, ritual life, political leadership, and decision-making. In fact, although we cannot generalize, there seems to be a correlation between female subordination and increasing social complexity (Costin 1996).

Women in state societies often have little power, prestige, and status, and are largely subordinated to their male counterparts (fathers and husbands) as well as to the patriarchal state (Ginsburg and Rapp 1995; Yuval-Davis 1993). Some archaeologists have been successful in demonstrating that the advent of state societies brought more workload, social strain, and less mobility for peasant women, although the requirements of a tributary state are likely to have largely affected men and children's lives as well (Brumfiel 1991; Hastorf 1991). Accordingly, the process of increasing centralization and social inequality would lead to a growing concern in regards to defining social identities.

In contexts where the elite controls craft production, the iconography on material items reflects a hegemonic discourse on the expected social identities and gender roles. Therefore, the examination of the iconography, in contrast to other lines of evidence for social inequality and political centralization, can be used to better understand patterns of sociopolitical complexity. Although ceramics iconography has been debated (Gomes 2001; Guapindaia 2001; Schaan 2001c), few attempts have been made to apply a gender perspective to iconographic studies in pre-Columbian Amazonia (Roosevelt 1988; Schaan 2001a).

In my opinion, the more prominent position of males in Santarém iconography is related both to the political centralization in the hands of chiefs and to male domination of public rituals, activities, and warfare. The rare effigy that features a female holding a bowl and wearing a *muiraquitã* (Gomes 2001:141), supposedly a prerogative of elite women (Boomert 1987; Gomes 2001), may mean that although gender hierarchy existed, it did not affect elite females in the same way. The clear separation between males and females as well as a relative separation between humans and animals (some hybridism still persisted in caryatid vessels) shows that the Santarém society was moving toward a greater valorization of humans over nature and males over females.

When comparing anthropomorphic representations in Santarém phase ceramics with those of the Marajoara phase, representations are much more realistic in the former than in the latter. I assume that these differences are related to sociopolitical organization and gender relations. In Marajó, power was probably more diffused, and in the hands of lineages, for hybrid figures, generally related to ancestors and the supernatural, were represented. Conversely, in Santarém phase ceramics, male and female figures are represented in rich details, as if specific individuals were being depicted. In both cases, however, the iconography represents the dominant caste and their ancestors: it is an iconography of prestige. I would say that in the Marajoara case, the anthropomorphic iconography suggests the political and religious power of the elite, whereas in Santarém phase ceramics, it represents their leaders.

## Tapajó-Konduri Relations

In examining the relations between the Tapajó and Konduri peoples, I examine three cultural variables: participation in the greenstone exchange network, similarities in the ceramic style, and similarities in settlement patterns.

In Mesoamerica, the exchange of exotic materials such as feathers and jade is believed to have played an important role in a "rewards system" through which rulers maintained their power and acquired more followers. According to this view, specific goals and strategies were behind the exchange of goods (Blanton et al. 1981). Some scholars criticize this position, arguing that this hypothesis assumes the existence of a central authority and central rule, which would not be true, for example, in the Mayan case (Freidel 1993). According to Freidel (1993), traders, politicians, farmers, and craftsmen in the cities used greenstones as currency, despite their intrinsic value related to their magical power and sacredness.

The dynamics of both the production and the exchange of greenstones follow a different pattern if we compare Mesoamerica and South America. In Mesoamerica, a variety of greenstone ornaments and objects were produced by specialists and used by the elite as part of a display of luxury, secrecy, and power. In South America, greenstone ornaments were probably used as "primitive valuables" (Sahlins 1972) or currency in a system that involved reciprocity, reinforcement of alliances, and exchange for other products. In Mesoamerica, greenstone symbolism was used to reinforce power within polities, through the control of the sources of raw materials and the sponsoring of specialists. In South America, greenstone symbolism was used to reinforce power within and between polities, through the control of exchange routes and esoteric knowledge (in the sense Helms [1979] uses the term). These two different sources of power are thought to have been important for the maintenance of a centralized power in chiefdoms and emergent states (Renfrew 1982).

The differential use of greenstones in the two regions—Mesoamerica and Costa Rica on the one hand, and Central and South America on the other—is probably related to how much access they had to jade sources. Lathrap (1973) proposed that the unequal access to raw materials due to environmental differences, coupled with the craft skills involved in producing localized specialties, led to complex indigenous trade networks, integrating a number of often hostile and ethnically diverse indigenous groups.

Greenstone sources are unknown south of Costa Rica, so they had to be obtained through trade. The large distances that separate Mesoamerica and South America, together with the fact that trading was not done directly, but probably counted on intermediation, can be seen as the main reason for the rarity of greenstone ornaments in northern South America and the Amazonian lowlands. Moreover, that scarcity most certainly elevated the value of the objects. Therefore, the appearance of these objects in the archaeological record should be interpreted in the exact dimension of what they represent. Jade ornaments were more rare in South America than in Mesoamerica, and thus meant power, alliances, contact with the external world, and mastery in trading.

The fact that greenstone pendants and statues have been found around the Nhamundá-Trombetas and Tapajós Rivers is illustrative of their relations. Both were possibly active participants in the greenstone trade, which affected not only the regional polities located in the Amazon basin but also other social formations, in a network that involved northern South America and possibly the Caribbean.

Further comparative studies are necessary to determine the exact trade routes. These studies should focus on identifying raw materials and comparing styles and carving techniques. These kinds of data would be extremely useful in defining alliances among complex societies in the late pre-Columbian period, and could be used to help explain the particular sociopolitical processes that led to the development and the fall of ancient chiefdoms and states.

One of the interesting motifs used throughout the Americas is the carving of frogs in small pendants. These frog-shaped pendants should be considered a leading piece in determining cultural connections between distant polities. This symbolism, which is probably related to water, fertility, agriculture, and procreation, shows a unifying cultural meaning that unites peoples living in tropical environments.

In the sixteenth and seventeenth centuries, natives constantly told Spanish and Portuguese voyagers that the greenstones were produced by the Amazons living in the north, far from the Trombetas and Nhamundá Rivers' mouths at the Amazon River. This tale was certainly a clever way to increase the value of those goods, which became highly desired and were owned by a small number of people. Greenstones were not only obtained from a distant place, but were also produced by magical processes, since their raw material was the mud which hardened out of contact with the air.

We do not have enough data to elaborate on the relationship between the Tapajó and the Konduri. However, their physical proximity, their similar ways of occupying and transforming landscapes, and the fact that they shared a similar pottery style suggest an involvement in a same political economy, maybe fitting into a peer polity interaction model (Renfrew 1986). The largest sites located in the present cities of Santarém and Oriximiná were sizable political centers where goods and people converged. More in-depth studies of critical sites in the region are needed to map workshops of specific goods (including cloth, lithic objects, and ceramics), fishing stations, agricultural lands, and secondary centers.

Some authors have sought to find connections between Konduri/Tapajó pottery and other ceramic styles elsewhere. Erland Nordenskiöld (1930) compared them to the ceramics found in Santo Domingo (Antilles), Panama, and Costa Rica. Helen Palmatary (1939, 1960) agreed with him, adding British Guyana, Venezuela, and southern U.S. mound sites. In the sixteenth century, the Antilles were occupied by Arawak groups, who have a diasporic distribution along the Amazonian periphery, including northern South America. There is no information on whether the Tapajó and the Konduri were Arawak or not, but some characteristics of their ceramics certainly point in this direction.

## Note

1. The idea that the ceramic firing process would produce the charcoal found in the soil has been suggested by Paulo Marcelo Paiva (personal communication, 2010).

*Chapter 5*

# Marks on the Earth: Territoriality and Memory

"Sculptured structures are bodies. Their matter, consisting of different materials, is variously formed. The forming of it happens by demarcation as setting up an inclosing and excluding border. Herewith, space comes into play" (Heidegger 1973).

## The Discovery of Enclosures in Acre: The Geoglyphs

Archaeological research in the Brazilian portion of western Amazonia began later than in other parts of the region, especially in the lower Amazon, which has been studied since the nineteenth century (see, for instance, Derby 1879; Ferreira Penna 1876, 1877; Hartt 1871). The first archaeological sites in Acre were recorded only in the late 1970s as part of an Amazonian surveying program (PRONAPABA) promoted by a team of scholars affiliated to the National Research Council (CNPq) and the Smithsonian Institution, which had previously developed a similar program for the remainder of the country (Dias 1995). In 1977, while surveying eastern Acre, a team of Brazilian archaeologists from Rio de Janeiro found sites containing potsherds and circular ditches, with accompanying embankments (Dias 1977). In a publication 11 years later, Dias and Carvalho (1988) described the eight circular ditches identified, with diameters varying between 50 and 120 meters; most of them were already partially destroyed by farming and ranching.

In the 1970s, the military government launched a series of projects promoting the colonization of Amazonia to integrate the region with the national economy. These projects included the construction of highways across the tropical forest from south to north and from east to west. Moreover, competition for restricted land in the south, along with the poverty caused by droughts in the northeast, drove farmers and refugees to the north, where the government guaranteed vast land plots. Upon arriving in Amazonia, immigrants cut down the forest to build their houses and plant their crops. In Acre, deforestation uncovered enormous ring ditches that for many centuries had been concealed by luxuriant tropical vegetation. Most farmers realized that such ditches were not naturally produced, but were most certainly excavated by humans. Their first thought was that the ditches were remnants of the Acrean revolution in the beginning of the twentieth century, the oldest historical fact they could recall. For archaeologists, however, the presence of potsherds on the surface was undeniable evidence that the earthworks were of pre-Hispanic origin.

What were these features built for? Dias and Carvalho doubted the ditch/berm complex had defensive purposes, since the berm was built in the outer perimeter of the ditch; instead, they believed an internal embankment would have be more effective, raising the position of the inhabitants vis-à-vis potential attackers. At that time, little publicity was given to the sites, and even the academic community was unaware of their discovery. During the 1970s and 1980s, Amazonian archaeology was dominated by cultural ecology, a paradigm that promoted the idea that cultural development would be similar throughout the whole basin. Villages were expected to be small, and vestiges would be composed mainly of potsherds, with some lithic artifacts. Field methodology consisted of collecting pottery samples for seriation, following the Ford method, and stratigraphy was considered too blurry to tell (Evans and Meggers 1965). Therefore, earthworks were largely ignored, and debate about their implications for social complexity avoided. As a result of his work in Acre, Dias (2006) assigned the ceramics to a local tradition called Quinari, related to *terra firme* horticulturalists, who lived in small sites, and produced small amounts of ceramics, mostly undecorated. At the Lobão site, they found large ceramic vessels containing human osteological remains, which they unfortunately did not describe. One type of decorated vessel was thought to be typical of the Quinari Tradition, although its archaeological context is uncertain; these were called *vasos caretas* (face vessels), as they have a cylindrical shape, with a globular mid-section displaying an anthropomorphic face (Figure 5.1). The ceramic technology used in the manufacture of face vessels is similar to that of the Quinari Tradition, so it is possible that these were related to the ditch builders.

The magnitude of the circular ditches was appreciated from above for the first time in 1988. Alceu Ranzi, a geographer and paleontologist who participated with Dias as an undergraduate student in the 1977 survey, was arriving in the city of Rio Branco on a flight from the city of Porto Velho when he spotted a double-ditched round figure (the Seu Chiquinho site) 1,500 meters below (Figure 5.2A). A few days later, with the support of the state government, Ranzi rented a small airplane and photographed the site. Thus, the first aerial image of what Ranzi referred to as a "geoglyph" made the local newspaper headlines. Another decade would pass until Ranzi was able to view another structure. In 1999, again flying from Porto Velho to Rio Branco, he saw a group of ditched structures (a circle, a rectangle, and a U-shaped enclosure) at the Colorada Ranch. Ranzi then realized he was in the presence of a regional phenomenon. In the following years, he recorded 20 other structures from small airplanes, visiting them on the ground whenever possible. He named

Figure 5.1. Face vessel (*vaso careta*; Museu da Borracha collection, Rio Branco, Acre). Photograph by Édison Caetano.

them geoglyphs, since they looked like marks on the ground, and could only be entirely appreciated from above.

It was a trip to Peru that inspired Ranzi to call the earthworks in Acre geoglyphs. Unfortunately, the term "geoglyph" inspires comparisons with the Nazca lines, which are a different phenomenon. In the Nazca desert, geometric and zoomorphic figures were shaped by the displacement of dark weathered rocks on the surface to expose a lighter subsurface. In Brazil and Bolivia, however, the "figures" were produced by the excavation of large continuous ditches forming circles, rectangles, hexagons, octagons, and other nongeometric shapes.

Looking for cooperation to investigate them further, Ranzi contacted Martti Pärssinen and his Finnish team, who in 2000 were conducting an archaeological research in Riberalta, a Bolivian province located some 100 kilometers from the border, where an Inca settlement and some ring ditches predating the settlement had been studied. During the visit to Acre, the group did a short nonintrusive survey and later organized the data and presented some hypotheses in a couple of scientific articles (Pärssinen et al. 2003; Ranzi 2003; Ranzi and Aguiar 2000, 2004). The articles were important in setting a base from which to create a research project and obtain funds for a first field season. I initially joined the project as the Brazilian archaeologist responsible for the legal permits.

In 2005, I led a multidisciplinary team that included Ranzi, some geographers, an ecologist, and an engineer from the local university and began a intensive survey in the area. In addition, some members of the team found out that they could easily identify the geoglyphs using free satellite imagery from Google Earth (Ranzi et al. 2007). In a few weeks, the number of known earthworks had tripled. Other features, such as

Figure 5.2. (A) Seu Chiquinho (photograph by Sérgio Vale); (B) Colorada Ranch, the first geoglyphs seen by A. Ranzi (photograph by Sanna Saunaluoma).

roads and small oval and lineal mounds, were also found. Based in the Geoprocessing Lab at the Federal University of Acre, researchers and students would spend hours, days, and weeks searching dozens of satellite images, covering all the areas with good images. After locating the sites at the lab, the team went to the field to take notes, photographs, and measurements, as well as to collect artifacts.

That same year, through a partnership between a Brazilian-Finnish team supported by CNPq, the Academy of Finland, and the state government, another project was planned to excavate some of the most emblematic sites (Pärssinen and Schaan 2005). Both projects were responsible for putting the state of Acre on the archaeological map, revealing one of the most intriguing archaeological phenomena in the last decade. A region that was once thought to be unsuitable for the settlements of pre-Columbian peoples was now the setting of ditches, berms, mounds, and roads, features that are more compatible with the occupation of complex societies.

Figure 5.2. Continued

## The Regional Configuration of Geoglyphs

The ditched enclosures known in Brazil as geoglyphs are found in the interfluves of the Acre, Iquiri, and Abunã Rivers (Figure 5.3), on the edges of plateaus. The first two rivers form the Purus River, whereas the Abunã River, which forms part of the border between Bolivia and Brazil, is a tributary of the Madeira River.

Each site is formed by a single enclosure or by a group of enclosures, which are sometimes associated with mounds and bermed roads. There are 190 sites in the state of Acre; these include 271 enclosures in the form of ditches with adjacent embankments as well as a few ditchless enclosures formed only by berms. Although we did not systematically study enclosures in the adjoining states, we have identified 20 structures in Rondônia and 10 in the state of Amazonas (Pärssinen et al. 2009) through satellite imagery. In Bolivia, 76 circular ditches have been studied by different archaeological

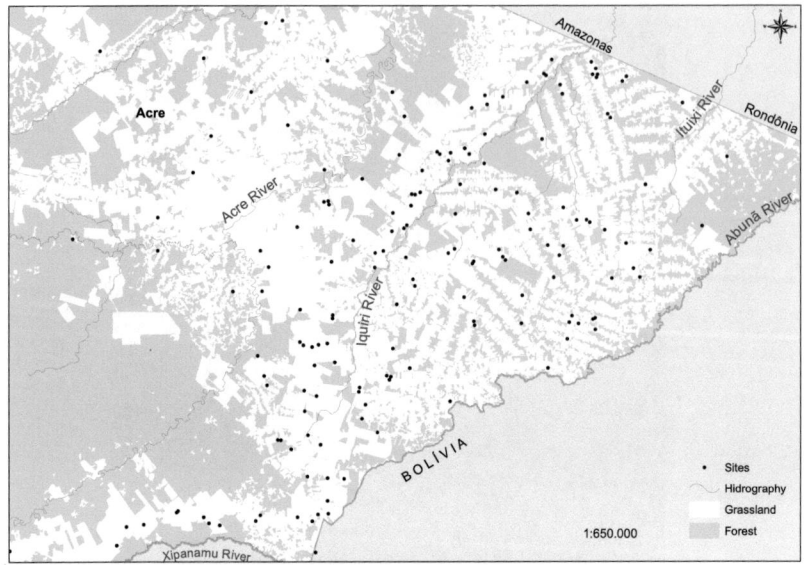

**Figure 5.3.** Location of the ditched enclosures (geoglyphs) in the upper Purus basin.

teams (Arnold and Prettol 1988; Calandra and Salceda 2004; Denevan 1963; Erickson 2010; Erickson et al. 2008; Saunaluoma 2010). In Acre, ditched enclosures are located on plateau tops and often at their edges (between 180 and 230 meters above sea level); in the state of Amazonas, they are located on the floodplains or at lower altitudes, whereas in Bolivia, they are found on river bluffs, forest islands, or floodplains.

The excavated ditches are generally 11 meters wide at the top, with depths varying between 1 and 4 meters in most cases. Today, these ditches are partially filled with eroded soil, which means that they used to be 1 to 2 meters deeper. Heavy erosion has also affected the external embankment, which is now between 0.5 and 2 meters high. In double-ditched enclosures, the ditch/berm complex may be as wide as 25 meters. The volume of dirt excavated for building a 100-meter-diameter circular enclosure, with a 3-meter-deep and 10-meter-wide ditch, for example, would be around 3,000 cubic meters. Assuming that a man could remove 1 cubic meter per day (using digging sticks and baskets), it would be necessary to have 3,000 persons per day to do the job, or 100 persons working for 30 days. Because workers generally have families, and need food and shelter, a larger population was probably involved in such activities, and for longer periods of time.

Prehistoric ditches are not new to archaeologists. In eastern Europe, they appear during the Neolithic period, enclosing villages, houses, and cemeteries. Neolithic ditched enclosures are rarely perfectly circular, nor are they always continuous, and were built with the probable intention of defining social spaces. In the literature, they are called enclosures or causewayed enclosures (Darvill and Thomas 2001; Oswald et al. 2001), because in some cases the trenches are not continuous, but are interrupted by causeways. Archaeologists have learned that causewayed enclosures have so much variability in terms of function, time period, and cultural content that the only characteristic they likely share are construction techniques (Calado 2009). The same seems to apply to western Amazonia.

The location of sites in Acre suggests defensive concerns because they are strategically situated in the landscape, with settlements built on the highest possible spot in a position of visual advantage in relation to those coming from the river valleys. If the forest did not exist during the period when the sites were occupied, the view would have been magnificent. It is possible that the placement of sites at higher elevations also had symbolic and religious significance. Distance from rivers indicate that navigation was not important. Distance to the nearest rivers vary between 1.5 and 8 kilometers. Water, however, was available year-round. All sites have good sources of spring water nearby.

In contrast to Bolivian sites, which are usually circular, oval, or form irregular structures, Brazilian ditches form rectangles, squares, and polygons. Two, three, or more ditched enclosures are sometimes built in conjunction or in proximity (with or without connecting roads), forming a complex. This fact notwithstanding, 59 percent of the sites contain a single enclosure without any other earthworks nearby, 27 percent of which are circles and 32 percent of which are rectangles or squares.

Twelve sites have double ditches: double squares with rounded edges (seven sites); double circles (two sites); double U-shaped ditches (one site); a circle inside a square (one site); and a square inside a circle (one site).

Site layouts vary, and 26 percent of the sites comprise a combination of two or more enclosures. Some of the existing patterns include:

(1) A circle and a square linked by a bermed road (three sites): Fazenda Califórnia, Rapirã, and Fazenda Atlântica (Figure 5.4A).
(2) Two squares linked by a bermed road (two sites): Fazenda Paraná and JD (Figure 5.4B).
(3) Two squares without any connection (four sites): Fazenda Três Meninas, Jacó Sá, Boa Sorte, and Santa Teresinha (Figure 5.4C).
(4) Two circles of different sizes with a common ditch in a tangential point, producing an anthropomorphic figure (three sites): Santa Izabel, Fazenda Soberana, and Fazenda São Paulo (Figure 5.4D).
(5) A ridged, ditchless rectangular court in between a bermed road and a ditched enclosure (three sites): Chinésio, Balneário Quinauá, and Fazenda Colorada (Figure 5.4E).

Figure 5.4. Examples of different site layouts: (A) Fazenda Atlântica (photograph by Sanna Saunaluoma); (B) Fazenda Paraná; (C) Santa Teresinha; (D) Santa Izabel; (E) Chinésio; (F) Gavião (photographs by Édison Caetano).

(6) Complex and unique combinations of ridged and ditched enclosures with roads and lineal mounds (five sites): Gavião, Hortigranjeira, Fazenda Colorada, Balneário Quinauá, and Fazenda Crixá (Figure 5.4F).

Sites of similar shapes and layouts have an interesting distribution across the region (Figure 5.5). Square and rectangular shapes dominate the northern part of the region; complex combinations are more usual in the center of the region, between latitudes 10°34' and 9°50' S, whereas circular sites are more usual in the south (Neves 2002; Schaan et al. 2007). Whether this reflects changes through time, different functions, or cultural differences between occupants, it is too early to tell. Available radiocarbon dates place the occupation of these sites between cal 2,275 ± 35 years BP (Severino Calazans site) and cal 1,195 ± 30 years BP (Jacó Sá site). Much earlier dates for the Severino Calazans and Fazenda São Paulo sites suggest that they were occupied before the construction of the ditches (Schaan 2010b). The Fazenda Colorada site yielded the

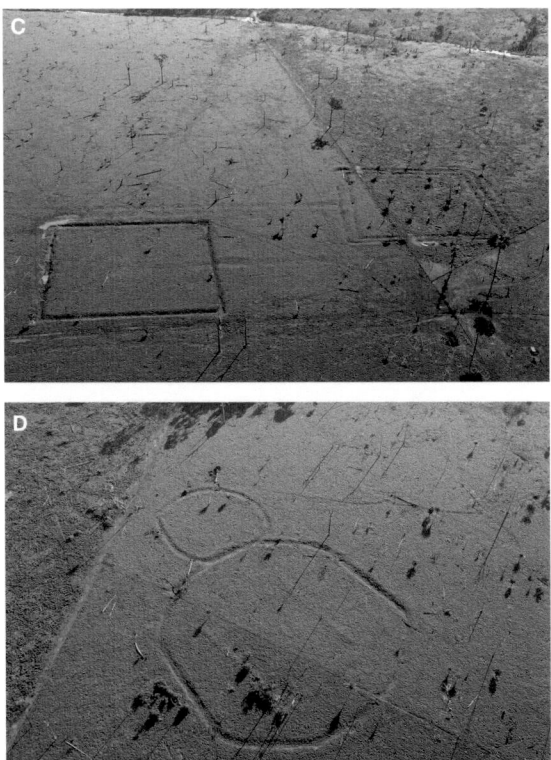

Figure 5.4. Continued

most recent date available, placing it at 750 ± 35 years BP (Pärssinen et al. 2003), which indicates that the "geoglyph culture" was mainly a first millennium phenomenon (Table 5.1). Occupations that predate the ditch builders are not well known. Ceramic sherds have been found outside ditches in some sites, eventually even 1,000 kilometers away (Ramal do Capatará site). Potsherds are also found in the berms, indicating that sherds were already in soil excavated from ditches and therefore were there before ditch construction. Nonditch sites here have low visibility because of the absence of highly modified soils (ADE has not been found in the state of Acre) and the low frequency of potsherds and other outstanding structures. Because of these characteristics, ceramic sites found here and in the Juruá River basin, in the western part of the state of Acre, are generally characterized as "camp sites" (Dias 1977).

## Mounds in Lowland South America

Geoglyphs are not only made of ditches and causeways; some are associated with other "positive" structures such as lineal mounds or straight,

Figure 5.4. Continued

short ridges that form a square, circular, or trapezoidal court. These earthworks have been heavily eroded by mechanized agriculture, and cannot be seen from satellites; in some cases, they are hard to spot even from the ground. We have identified them mostly through aerial survey during early morning hours, when the sunlight forms nice and unambiguous shades. I believe that two types of positive structures are important. The first is a square or trapezoidal bermed enclosure attached to a ditched enclosure (e.g., the Chinésio [Figure 5.4E], Balneário Quinauá, and Fazenda Colorada [Figure 5.2B] sites), which is also connected to a straight, long, bermed road. The second is a type consisting of several small mounds arranged around a circular opening or plaza; this is usually associated with a neighboring ditched enclosure, as in the case of the Fazenda Iquiri site (Figure 5.6). At the Coqueiral site, small mounds arranged in a circle had abundant ceramics on their surface. The pottery displayed different forms (miniatures and a spindle whorl) and decorations (including white incisions and polychromy), contrasting with the plain pottery that characterizes most of the geoglyph sites. In both cases, the mounds are no higher than 1.5 meters, with varying widths

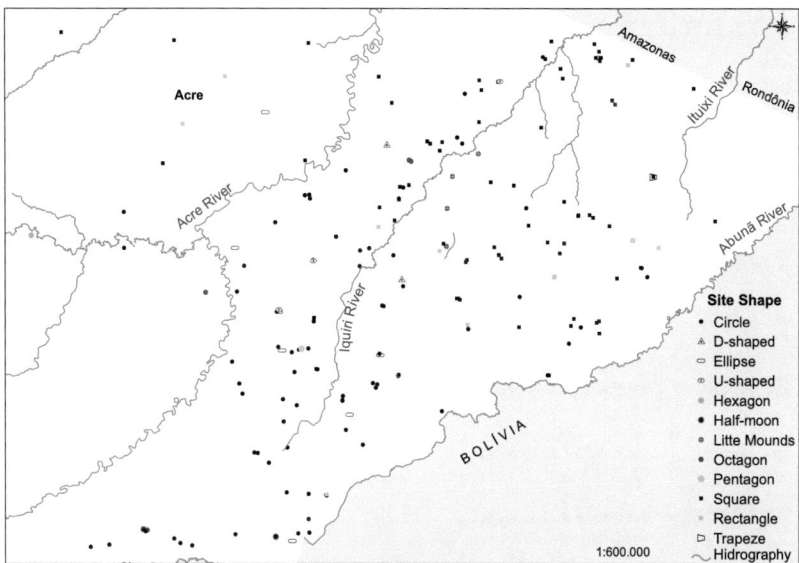

Figure 5.5.  Spatial distribution of site shapes.

(2–6 meters) and lengths (10–20 meters). None of the mounds has been excavated yet. It is unclear whether they were formed by the collapse of houses or if they are trash middens, and we do not know if they were built in single or multiple episodes.

Mounds are a common archaeological feature. Small mounds can be formed by either natural refuse accumulation or intentional construction; they can develop over a long period of time or be built expeditiously as part of a single event. They can also be a product of simple societies or represent a monumental construction effort related to the dynamics of regional integration between more complex societies. Many different types of mounds have been identified in the South American lowlands. The Cerritos of coastal Uruguay and Brazil are an interesting example (López-Mazz 2001). The size of these structures varies from 20 to 40 meters in diameter and 0.5 to 7 meters in height. They appeared as early as 4,000 years BP, and were the product of hunter-fisher-gather populations who may have been marking their economic territories. Through time, the Cerritos changed in function, as their users became more sedentary. During this period, ca. 2,500 years BP, the Cerritos took on a greater funerary and ceremonial role, either as collective cemeteries for villagers of both genders of all ages or as funerary tombs for a single leader. By 1,000 years BP, domestic structures were connected to cemeteries. Multiple episodes of reoccupation and remodeling of the mounds have been documented, as well as moments of greater regional integration (López-Mazz 2001). This example indicates that certain types of earthworks can develop across a

Table 5.1. Radiocarbon dates

| Lab number | Site, provenience | Depth, cm | Conventional $C^{14}$, years BP | Calibrated $C^{14}$ (2 sigma) |
| --- | --- | --- | --- | --- |
| Ua-37237 | Severino Calazans, Unit 3 | 50 | 3,990 ± 40 | 2577–2345 BC |
| Ua-37238 | Severino Calazans, Unit 5 | 45 | 2,915 ± 35 | 1211–942 BC |
| Ua-37265 | Severino Calazans, Unit 6B | 50–60 | 2,275 ± 35 | 171 BC–AD 25 |
| Ua-37264 | Severino Calazans, Unit 3 | 20–30 | 2,050 ± 35 | 159 BC–AD 57 |
| Ua-37235 | Fazenda Colorada, Unit 7 | 150–160 | 1,865 ± 65 | AD 25–342 |
| Ua-37256 | Fazenda Colorada, Unit 12 | 90 | 1,820 ± 30 | AD 131–326 |
| Ua-37567 | Fazenda Colorada, Unit 9 | 218 | 1,775 ± 35 | AD 144–394 |
| Ua-37259 | Jacó Sá, Unit IVB | 10–20 | 1,485 ± 35 | AD 545–650 |
| Ua-37236 | Fazenda Colorada, Unit 10 | 70–80 | 1,340 ± 35 | AD 653–773 |
| Ua-37255 | Fazenda Colorada, Unit 10 | 67 | 1,275 ± 30 | AD 675–869 |
| Ua-37258 | Jacó Sá, Unit 8 | 80–90 | 1,205 ± 30 | AD 730–962 |
| Ua-37257 | Jacó Sá, Unit 1 | 47 | 1,195 ± 30 | AD 776–966 |
| Hela-616 | Fazenda Colorada | 25 | 750 ± 35 | AD 1244–1378 |

region with greater significance over time, although their functions may vary considerably.

Small funerary mounds are also found inside earthen geometric enclosures in the highlands (Altiplano) of southern Brazil, between the states of Rio Grande do Sul and Santa Catarina (Iriarte et al. 2010). These enclosures are surrounded by 3- to 6-meter embankments enclosing a diameter of up to 180 meters and forming a circle or ellipse. Some of these enclosures have a small mound inside measuring 1.5 to 20 meters in diameter and 0.7 to 3 meters in height. Enclosures are located on hilltops, with an excellent view of the surroundings. Excavations have shown that these mounds were built in a single episode on top of either a single or multiple burials. Radiocarbon dates indicate that this monumental tradition, related to the Taquara-Itararé Tradition, grew in importance after AD 1000. Similar to geoglyphs in western Amazonia, the entrances to some of these bermed enclosures were made through wide (18-meter) roads that extended for more than 400 meters, apparently built with the purpose of directing people to their interior.

Figure 5.6. Fazenda Iquiri site: Mounds and ditched enclosure (aerial photograph by Diego Gurgel).

As in the case of the Cerritos, these enclosures are related to Macro-Gê language groups, whose funerary rites are supported by ethnohistorical information. According to these, the death of a chief would require the presence of secondary chiefs, who would come and help build the funerary mound. This was part of a long ritual that ended with the ascent of the chief's son to the position of leadership (Iriarte et al. 2010; López-Mazz 2001).

In the central Amazon region, a group of small mounds (up to 1.5 meters high) disposed as a half-circle was found inside the 400-square-meter Hatahara site. These mounds were formed by earth (mostly ADE) as well as large ceramic sherds, which were used as construction material. Excavators found burials underneath one of the mounds (Machado 2005). In this case, these mounds probably were intentionally built as termination rituals, possibly related to the death of a leader, although their circular configuration opens the possibility that other factors might have influenced their construction.

Along the upper Xingu River, in a region situated in the southern Amazon periphery, Michael Heckenberger (2010) has studied and described large complexes of interrelated earthworks composed of ring and semilunar ditches in and outside settlements, as well as straight roads connecting nearby settlements. Villages typically have a large central plaza (between 120 and 175 meters in diameter) surrounded by houses and backyard trash middens. In this case, mounds are trash middens, lineal curbs alongside roads, or embankments along defensive ditches.

Although Heckenberger's study focuses on the development of regional monumentality in what today is a Carib-speaking Kuikuru territory, he believes that the upper Xingu communities were influenced by Arawak groups, who arrived around AD 1000 (Heckenberger 2005, 2010).

## Figuring Geoglyphs

One question that has driven our regional study is the search for spatial patterns that would make sense of the configuration of site distribution across the landscape. Archaeologists have used models from geography to make inferences about regional interaction (Crumley 1979; Johnson 1977). The researcher defines a region and a scale based on the limits of an observable phenomenon (Crumley 1979). Although the geoglyphs are a regional phenomenon, we do not know if they were all contemporaneous (probably not); thus, we cannot define a proper scale for study, although we can experiment with a regional analysis.

When we look at the regional distribution of geoglyphs, the absence of clear clusters that would indicate the existence of one or more centralized systems is noticeable; however, the relationship between site distribution and size suggests a pattern. Figure 5.7 shows the spatial distribution of sites normalized by area (including clusters of sites containing two or more enclosures that are considered as a single site). Large sites are distributed somewhat regularly across the region, which might indicate several discrete settlement systems or a "heterarchical" (Crumley 1979) regional organization. Crumley (1979) defines heterarchy as the

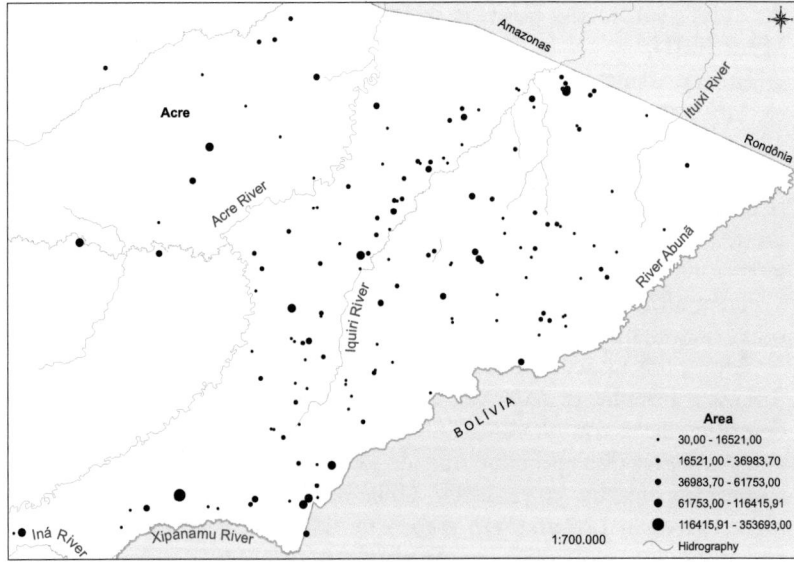

Figure 5.7. Regional distribution of sites, normalized by area.

different ways in which elements might be ranked to one another, a relationship that varies through time and space. On the map, the largest site is an octagon enclosure associated with a half-moon ditch located 800 meters away; the second largest site in size is an ellipse; and the third, a group of four enclosures: an ellipse, a circle, a rectangle, and a U-shaped enclosure. The five largest sites are located in the south of the region, not far from the Bolivian border.

We should acknowledge that this site distribution might be a product of sample bias; despite our efforts to cover the entire deforested area (Figure 5.3), earthworks located inside the forest cannot be identified by our aerial survey methods.

## Excavating Geoglyphs

Between 2007 and 2010, during different field seasons, we test-excavated eight sites, recovering few cultural features or artifacts. Martti Pärssinen directed the excavations at the Fazenda Colorada site (Figure 5.8). An excavation at the base of the outer ditch surrounding the U-shaped structure suggested an original ditch depth (the thickness of the cultural soil) of 2.6 meters measured from the ground surface, and 4.2 meters from the top of the outer embankment. Another excavation at the base of the inner ditch indicated an initial depth of 3.4 meters measured from the ground surface, and 5 meters from the top of the embankment. In this unit, Pärssinen observed a horizontal 20-centimeter-wide brown stratum at the 260- to 280-centimeter below surface level, which he believed to be decomposed timber, placed as a step to maintain pressure on the steep bermed structure (Schaan et al. n.d.).

While excavating the outer berm at the Fazenda São Paulo site, we identified a 25-centimeter-diameter and 1.3-meter-long post mold, filled with brown soil. Unfortunately, we did not have time to extend the excavation along the berms to find any similar features that would suggest a palisade.

At the Fazenda Atlântica site, Sanna Saunaluoma found remains of adobe near the main entrance to the enclosure, which she interpreted as the remains of a wattle-and-daub house.

At several sites, we have found ceramics at different depths in ditch berms, at ditch bases, and in the fill of outer embankments, indicating that these sites were occupied before and during construction. Despite these finds, the amount of cultural materials recovered during excavations was disappointing. For example, in two of the five circular ditches at the Ramal do Capatará site, we excavated 14 1-meter by 1-meter units, along with a 26-meter-long, 1.35-meter-wide, and 1.7-meter-deep trench cutting the first ditched enclosure transversally; in addition, we performed 120, 25-centimeter-wide and 1-meter-deep soil cores. Despite the ample sampling, we only recovered 1,223 ceramic sherds. In most ADE sites, this number of sherds would be recovered from a single 1-meter by 1-meter unit.

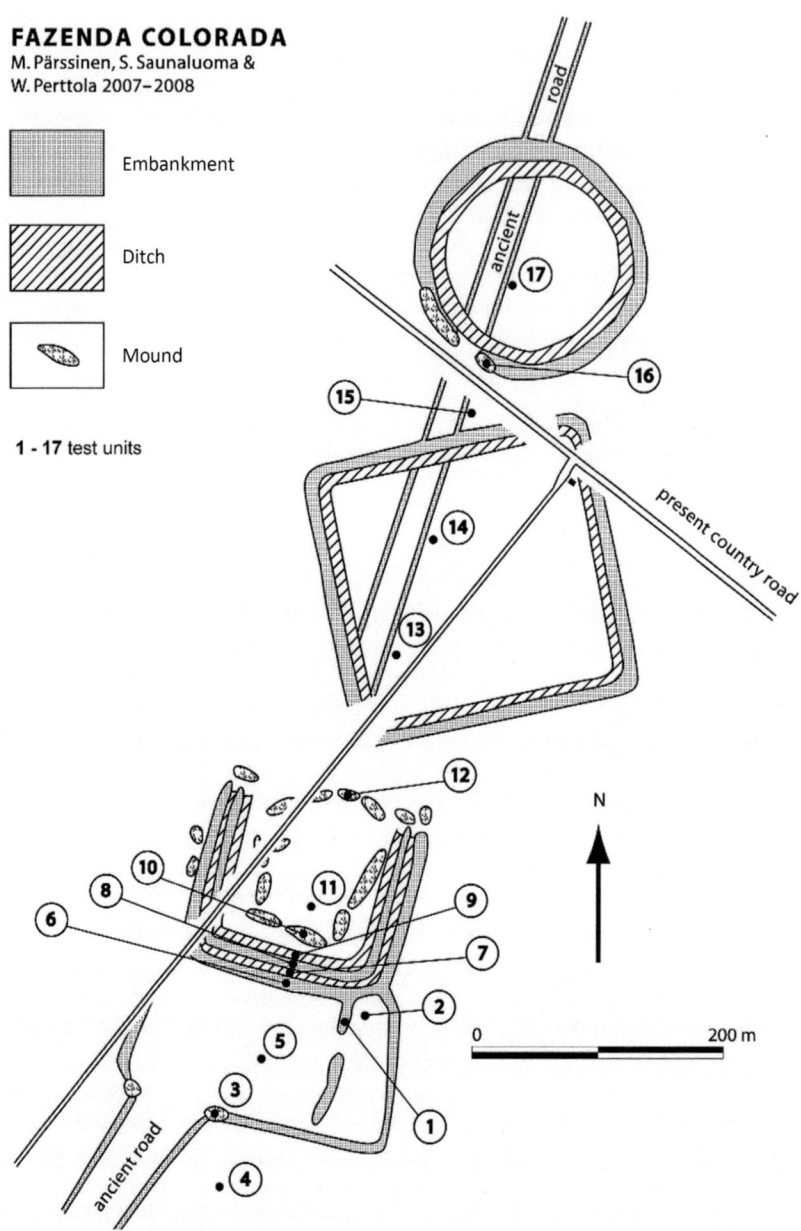

Figure 5.8. Fazenda Colorada plan (by Martti Pärssinen).

The enormous communal effort assumed for ditch excavation is not reflected in the quantity of ceramics found in each site. Ground surveys recovered few surface artifacts; at some sites, nothing was found.

Even when found during ditch and berm excavations, ceramics are not as numerous as expected if these were village sites. Alceu Ranzi, Roberto Feres, and Foster Brown had estimated that 60,000 people inhabited the sites, assuming the ditches were village enclosures (Ranzi et al. 2007). However, the meager presence of artifacts suggests that habitation may not have been the sites' primary function.

Circular villages are the cultural signature of many Amazonian groups, such as the Arawak, who are known agents of landscape modification, through the excavation of defensive ditches, as is the case in the upper Xingu River (Heckenberger 2005; Heckenberger et al. 2003). Dense populations living for long periods in the same area should produce impressive amounts of cultural materials. In Amazonia, dense settlements have resulted in the production of ADE, discussed in Chapter 4. These modified soils, however, are absent in the upper Purus River sites. The small habitation and campsites recorded during surveys in the 1970s do not contain ADE. Is this the result of short-term settlements? Could a small population have built the enclosures to use for short periods of time? Were the sites simply used as gathering places during certain times of the year? If so, where did the population live the rest of the time? A possible explanation for this apparent paradox—monumental construction with no signs of habitation—would be that these earth-moving activities had a religious purpose, a reason compelling enough to require enormous sacrifices from small, dispersed communities who would meet for short periods at set times of the year for religious cults and feasts.

This is in accordance with the monumentality of the enclosures, with enclosed ditch spaces as large as 28 hectares. These earthworks would be like cathedrals (Heckenberger, personal communication, 2007), monuments meant to show how small human beings are compared with a given deity.

The wide and deep ditches that form the geoglyphs were designed to enclose an open, flat area that could have been ideal for different types of communal gatherings (religious festivities, trading, rituals, and annual crop feasts). Large bermed roads led people coming from different places to the ditched enclosures. These straight roads are 18 to 25 meters wide, some of them extending for more than a kilometer. Some roads come from the rivers. Some roads connect two enclosures (Pärssinen et al. 2009). Earthworks inside the enclosure's central opening or plaza, such as small mounds and smaller ditched enclosures, could have served as boundaries for different groups of people.

## Looking at Artifacts

Despite the scarcity of samples, pottery found during surveys and excavations have allowed us to define local ceramic industries. The potsherds found during excavations are fragile, since they are embedded in compact sediments. Sherds are cleaned at the lab with a dry, soft brush, which helps maintain their original coat. Most ceramics are tempered with a combination of grog and *caraipé*. About 8 percent of the sherds collected from the surface are decorated, but only 4 percent of those recovered from excavations are decorated. White slip, incisions, and white, red, or black paint are the preferred decorative techniques. As a general rule, incisions are carelessly executed, as in this vessel found at the Fazenda Atlântica site (Figure 5.9). Although poor in decoration, the pottery is well fired.

Vessel shape varies, but most of the sherds belonged to utilitarian vessels, such as pots, plates, griddles, and robust vessels probably used for grinding. We have noticed everted, thickened rims with narrow lips at a couple of sites.

Few lithic artifacts have been found, most of which are worn stone axes that could have been used for ditch excavation. These await further study.

## Early Andean Contacts and Spanish Explorations in Western Amazonia

Western Amazonia has been an object of mystery and wonder since Inca times. According to Garcilaso de la Vega, the famous sixteenth-century Peruvian historian, in the mid-fifteenth century, Yupanqui—the first Inca to conquer new tribes and lands—sent 10,000 men on a military expedition from Cuzco to the east, on a quest to find the land of the Musu (Mojos), whom they regarded as advanced peoples and potential Inca allies. Only a few Inca soldiers, however, made it to Musu. According to Garcilaso, most died along the way, killed by the natives who had their territories invaded, or died of starvation while wandering in an unknown land. The few that arrived safely in Mojos stayed there, married local

Figure 5.9.   Fazenda Atlântica vessel excavated by Sanna Saunaluoma in 2007 and restored by Raimundo Teodório (photograph by Denise Schaan).

women, had children, and never returned to Cuzco. According to Gott (1993), this tale fueled the legend of El Dorado; however, others think this story might include some truth (Armentía 1887).

During early colonial times, traveling from Cuzco to the Bolivian lowlands required accessing the source of the Madre de Diós River, which begins in the Andes and empties into the Beni River. The province of Mojos is situated between the Beni and the Mamoré Rivers, which together form the Madeira River, one of the largest tributaries of the Amazon (Figure 1.1). The sources of three important tributaries of the upper Amazon—the Ucayali, the Purus, and the Madeira Rivers—are located near the Andes and not far from the Purus basin headwater. People probably traveled back and forth between the Andes and the Amazon for many centuries, and this path was likely used for the exchange of goods and especially ideas that mutually influenced cultural developments in these two broad regions.

Donald Lathrap (1970a) once proposed that the eastern Andes and the western Amazonia were part of a same cultural area, and that they maintained a strong interaction, since the Andean highland Chavín culture emerged in the last millennium BC. The depiction of particular species of the forest fauna—jaguars, eagles, and the caiman—in Chavín iconography suggests close connection to the tropical forest: "if the forgers of the Chavín art were not inhabitants of the Amazonian lowlands, at least they were notably familiar with its fauna, which suggests a great proximity with this ecological zone" (Lathrap 1970a:239). In his in-depth analysis of the Tello Obelisk as a representation of the cosmos at the site of Chavín de Huantar, the Chavín ceremonial center (2,900–2,500 years BP), Lathrap not only highlights its iconographic connections to the tropical forest, citing the pivotal role of the caiman (as "master of the fishes") and the depiction of jungle plants (manioc leaves), but he also proposes that the obelisk itself could have been carved down at the Marañon River drainage, possibly in Yauya, and later brought to the site (Lathrap 1985:249). Art historians tend to agree with Lathrap about the strong influences of the tropical forest on the Chavín art style (Stone-Miller 1995).

More than a millennium later, the Andean people's interest in the tropical forest was revived through the expansion of the impressive Inca Empire. In less than a century, through policies of alliance and war (when necessary), Tawantinsuyu, the Inca Empire, expanded to encompass northwestern Argentina, central Bolivia, the northern half of Chile, northern Ecuador, and southern Colombia. Historical and archaeological evidence of Inca strategies in the tropical forest has been documented by a team of Finnish-Bolivian archaeologists, who located a number of fifteenth-century Inca fortifications that expanded the outer limits of the empire to the east, reaching the province of Riberalta, located at the confluence of the Madre de Diós and the Beni Rivers, in the Bolivian lowlands (Pärssinen et al. 2003). According to Pärssinen and colleagues, this system of fortifications was part of the Inca strategy of pushing their empire further through invasions, sometimes not peacefully, especially when faced with peoples that were not organized around *señorios* or whose leaders would not gracefully submit to Inca rule. Archaeological research at Las Piedras, the Inca fortress at Riberalta, indicated that the site was densely populated; among the potsherds collected, there were a few Inca-style ceramics in an assemblage that was characterized mostly by Amazonian pottery, some of which resembled the Sivia style of the Peruvian rainforest (described by Raymond 1995). The site was positively dated to the mid-fifteenth century, and it was certainly occupied before the Inca presence. Archaeologists believe the site was taken during the Topa Inca expedition to the region documented by Garcilaso. However, there are no data on the nature of the relationship established between the Inca and local inhabitants (Pärssinen et al. 2003).

Additional possible material evidence for contact between the Andes and the lowlands is a road, mentioned in historic chronicles, but unknown to archaeologists. Missionaries traveling to the Madre de Diós region in the nineteenth century reportedly found vestiges of a 25-foot-wide Inca road from Cuzco to the Takana territory, located between the Madidi and the Mayu-Tata (Madre de Diós) Rivers, near the Mission of Ysiama (Armentía 1887).

Under Spanish colonial rule, several explorers were sent to Mojos with the task of acquainting themselves with that part of the territory and subjecting it to Spanish rule. Many of those expeditions, due to unknown territory and the crossing of vast mountain ranges, were unsuccessful, since travelers often died of hunger, disease, and native ambushes. Nonetheless, news brought back from the few surviving explorers to the governments in Cuzco and Santa Cruz stimulated new expeditions: "there is a large number of well-dressed people there, rich in gold and silver and cattle, with fertile lands," the governor of Santa Cruz believed around 1580, when he decided to send a new expedition to Mojos (Gott 1993). Despite all efforts, however, Mojos remained isolated for a century, although periodic encounters continued to fuel much speculation over the rich lands and civilized people who lived in the Bolivian lowlands. In 1597, missionaries obtained official permission to establish missions in the Mojos, but their first attempts were aborted when natives killed the priests.

Beginning in 1764, several Franciscan expeditions departed from La Paz and sailed down the Beni River to teach the Gospel to the Araona and Toromona settled along this important river. Many were unsuccessful due to ecological conditions and the lack of tolerance the native populations in this region had for priests. Detailed information about these missions is lacking. Father Nicholas de Armentía, who traveled to the region in 1881, explains that mission records were often destroyed by fires, since churches and missionary residences were always established in palm-leaf huts. During the nearly six years Armentía lived among the Takana peoples (especially the Araona and the Tomorona) and the Pacaguara (Pano), he visited many villages along the Madre de Dios, the Orton (Tahuamanu), the Purus, and the Acre Rivers. His written records are considered the most reliable source of information on the indigenous way of life in this important region, although the customs had already been affected at that point by the consequences of European incursions in the area. Our interest in his account lies in what we can learn about the peoples whose ancestors were likely responsible for the earthworks that are the focus of this chapter. I will return to this later.

### The Archaeology of Baures and Mojos

Between 1908 and 1914, Swedish anthropologist Erland Nordenskiöld visited the savannas of northeastern Bolivia between the Beni and Guaporé Rivers, describing for the first time several types of earthworks built as indigenous responses to seasonal floods in a region known as Llanos de Mojos (Denevan 2009; Nordenskiöld 2009). These open, seasonally inundated savannas encompass about half of the Department of Beni. In 1916, Nordenskiöld stated that one cannot fully understand that land without visiting it both in the rainy and the dry season, which is something I can truly appreciate, since I have said the same thing about the Marajó Island savannas (Schaan 2004). According to Nordenskiöld, flood intensity varies, some areas at Llanos de Mojos are more flooded than others, and the amount of high ground that remains free from flooding varies across the region. Some of the places high enough to avoid flooding are covered with forest, and are called *islas* for that reason. Nordenskiöld explains that the Andean rivers bring massive amounts of mud and debris into the Bolivian lowlands, leveling the land and making it fertile. According to historical accounts, especially those of the Jesuits, Arawak-speaking people built large towns and developed a highly productive agriculture.

In 1791, Father Francisco Eder mentioned the existence of earthen bridges (causeways) used by pedestrians and canals that could be navigated in canoes during the wet season, transporting maize and other goods built by the local inhabitants (Nordenskiöld 2009:217). Although this area was named after the Mojo, the canals and causeways were likely built by the Baures, an Arawak-speaking group who lived between the Blanco and San Martin Rivers when the Jesuits penetrated the savannas (Nordenskiöld 2009). Nordenskiöld described one of the canals that connected the Mamoré and Ipumpuru Rivers—which he himself navigated—as 2 kilometers long, 6 to 7 meters wide, and 3 to 4 meters deep. He thought that the idea of building causeways could have been imported from somewhere else, like the *tesos* of Marajó Island. These societies were probably contemporary, however, and the earthworks are different, although the problems they were meant to solve are similar. I believe that although the many different types of earthworks present at Llanos de Mojos relate to different practices, such as agriculture, transportation, and fishing, only fishing was fully explored in Marajó. As for the mounds, Nordenskiöld believed that people would not have gone through the trouble of building elevated areas simply for residential purposes, as they could have used pine dwellings instead; therefore, according to him, these elevated areas were designed for planting crops and burying the dead (Nordenskiöld 2009:219). He also noticed that some of the mounds were higher than necessary, pointing also to the existence of borrow pits nearby as testimonies of earthmoving. Dams built for storing water during the dry season were another type of earthwork described by Nordenskiöld; these were built for storing water for the dry season.

Inspired by Nordenskiöld, George Plafker (1963) and William Denevan (1966) documented additional earthworks in the area. Whereas Nordenskiöld described a few earthworks from his surveys and excavations, Plafker had a broader assessment of the region's raised fields, causeways, and canals, which he observed from aerial photographs. At a time when archaeologists were studying ceramic cultures in other parts of the basin, debating whether Amazonians' native technologies were capable of feeding large populations and supporting higher cultures, Denevan was using a landscape approach to study the region's archaeology, describing an environment that was highly manipulated and transformed by societies whose agriculture was probably more advanced than the slash-and-burn commonly believed to be the standard practice in Amazonia (Denevan 1992, 2001; Plafker 1963). Cultural ecologists thought that complex societies in the Llanos de Mojos were the result of Andean migrations to the lowlands, so that the incredible earthworks of the Llanos de Mojos were not an exception to the general rule that the tropical forest had not supported higher cultures. The cultural features Denevan and later Erickson studied in the Llanos de Mojos are now found to be common also in other parts of the Amazon basin. Earthworks had long been found in Marajó as well. These two archaeological regions, located on the Amazonian periphery, were studied early on because they are open savanna environment and thus exposed to western eyes. Other impressive and unknown pre-Columbian earthworks probably exist under the forest across the Amazon basin.

Denevan described three types of drained fields in the Llanos de Mojos: "(1) furrowed fields, where ditches had been dug to provide drainage; (2) raised fields, where earth had been piled up to form low, widely spaced rectangular platforms; and (3) mound fields consisting of regular spaced circular mounds" (Denevan 1963:540). The possibility of observing these features from airplanes and taking aerial photographs has made all the difference in this type of archaeological landscape, since many of these fields are low and leveled by erosion and sedimentation processes, and therefore are hard to see from the ground in most cases. Their extent, however, is impressive: Denevan (2001:246) estimated that at least 35,000 drained fields existed in western Beni. Denevan also described other important features: causeways and canals made for transportation across flooded lands, habitation mounds (with potsherds and burial urns), and circular ditches (presumably for defensive purposes). To him, the earthworks, which were likely built over many centuries by Arawak or Arawak-influenced populations, were a product of highly stratified societies, probably chiefdoms.

Denevan focused on the use of earthworks for agriculture, given the poor quality of savanna soils and the need for elevated terrain due to flooding, a study that was continued by Clark Erickson, who also did some experimental work with raised fields in the Llanos de Mojos (1980, 1995, 2000, 2001, 2006; Figure 5.10). In addition, Erickson described special forms of earthmoving related to fish farming (see Chapter 2), and has recently deepened his research to the study of ring ditches (Erickson 2010; Erickson et al. 2008), an interest shared by other archaeologists working in the area (Arnold and Prettol 1988; Saunaluoma 2010; Walker 2008).

Ditched enclosures in northern Bolivia are generally called ring ditches or *zanjas circundantes* because in most cases they have a roughly circular shape, although oval, octagonal, D-shaped, lineal, incomplete, and irregular rings have also been reported. They occur over a vast region, from the city of Riberalta, at the confluence of the Madre de Dios and Beni Rivers, some 80 kilometers from Brazil (Arnold and Prettol 1988; Saunaluoma 2010), to northeastern Mojos, east of the Mamoré River (Denevan 1966; Erickson et al. 2008). Denevan (1963) was the first to describe ring ditches in Llanos de Mojos in detail, but research efforts specifically directed at these earthworks are more recent in the region. Ring ditches in Acre and Bolivia share similar construction techniques, but geometry seems to be more important in Acre. Moreover, in Bolivia, ring ditches are sometimes associated with other types of earthworks, such as raised fields and mounds.

In central Mojos, Walker (2008) studied four ring ditches in forest islands surrounding mounds. These mounds were partially built with dirt excavated from the ditches, but further accretion brought about by subsequent occupations is necessary to account for present mound height. He also found extensive raised fields associated with three of the

Figure 5.10. Beni raised fields, Bolivian Amazonia (aerial photograph courtesy of Clark Erickson).

ring ditches, which he believes had hydraulic functions, such as draining and conserving water. Walker finds it difficult to think of the Yacuma ring ditches as barriers to protect against intruders, unless they were coupled with palisades, given their modest depth and width.

Near the mouth of the Beni River, Arnold and Prettol (1988) excavated the Tumichucua site, a complex of ditches shaped as straight lines, circles, and semicircles, located between the river and an oxbow lake, covering an area of approximately 1.25 square kilometers. Inside one of the ditches, they found two hardened clay balls they believed could have been used to anchor sharpened stakes for defense (Arnold and Prettol 1988:461). Like Walker, they also felt that the ditches could have been used to retain water, given their connection to the river and the lake. Arnold and Prettol stated that the earthworks at the Tumichucua site did not involve a massive workforce like the ones at Mojos, but their construction still implies a population larger than the groups historically known to have occupied most tropical environments.

Sanna Saunaluoma's research (2010) in the Riberalta region included new excavations at the Tumichucua site, as well as six other sites in the vicinity, all roughly circular or semicircular and lineal ditches. These sites are located near the multicomponent site of Las Piedras, discussed earlier in this chapter. The most recent earthworks were contemporary with the fortress, dated after AD 1300. Saunaluoma considers these ditches to be enclosed spaces used to mark occupational areas and serve both practical and symbolic purposes, protecting occupants from natural and supernatural forces. The sites in this region were dated to two separate periods: 2,100 years BP to 1,600 years BP, and after AD 1200. Saunaluoma believes that other sites belonging to the intervening period are probably located farther away from the Beni and Madre de Diós Rivers, given their active, meandering nature.

Regional chronology for the simple ceramics found in ring ditch sites is lacking. The amount of sherds recovered is usually quite small. Saunaluoma indicates similarities between the Riberalta ceramics and those of the upper Purus River, in Acre, and the upper Ucayali River. In two of the sites, however, she has found corrugated ceramics that are probably related to Guarani groups, who occupied that area in the seventeenth century.

Erickson and colleagues (Erickson 2010; Erickson et al. 2008) have identified and mapped up to 60 ring ditches in the region around Baures or the province of Iténez, in Bolivia. Their similarity of the geoglyphs to Acrean geoglyphs is impressive; ditch width is about 11 meters, and the diameter of the enclosures varies between 150 and 350 meters. Some sites have multiple concentric ditches (Erickson 2006). Erickson and colleagues (Erickson et al. 2008) believe that the rings had multiple functions such as "defense, settlements, elite residences, land and resource markers, animal traps, cemeteries, water management features, and/or public ceremonial spaces" (2010:620). Defensive function is supported by ethnohistorical reports. Father Eder noted that the Baure were in constant state of war against their neighbors, especially the Guarayo, who were described as violent cannibals. Moreover, ditches were steep and located on high terrain, providing good visual control of the savannas (Erickson et al. 2008:77). Different from other earthworks in the region, which are usually located in the flooded savannas, *zanjas* are more often found in forest islands.

Walker (2008) states that ring ditches cannot be attributed solely to Arawak language groups, since ethnohistorical accounts place six language groups in northern Bolivia, including two Arawak (Mojos and Baure) and four other isolated languages. The presence of ring ditches in the traditional region of the Movina in central Mojos undermines the exclusive association with Arawakan groups, despite the fact that circular ditches have long been described as features related to the Arawak culture in the upper Xingu River (Heckenberger 2005). An Arawakan origin for the ring ditches is supported by an Arawakan myth of the upper Xingu River according to which the cultural hero Viti-Viti was responsible for building deep, long, semicircular ditches to enclose villages for his people (Villas-Boas and Villas-Boas 1985:143).

The time span for ring ditch sites extends over two millennia, encompassing a vast, ecologically heterogeneous area of western Amazonia. Historical information indicates different linguistic groups occupied the area. Archaeology shows different types of earthworks with different functions. These sites were likely the product of more than one ethnic group; through the centuries, they were built and reoccupied, with different functions over time. As a construction technique, ditches had a long life, representing a unique feature in a region where population migration and conflicts required the defense and enclosing of settlements as well as the creation of ceremonial centers and gathering places.

### Occupation of the Upper Purus River Basin

Although the Spaniards were aware that significant populations inhabited the western Amazonia by the sixteenth and seventeenth centuries, nothing was said about the peoples who lived in the upper Purus River basin. The Purus River, for most of its extension, flows through an alluvial plain with occasional high clay bluffs. This is a river that has uncountable dramatic bends as well as a very dynamic history of course change, forming innumerable oxbow lakes along its path. The river itself, with the many products it had to offer the conquerors and the substantial population it supported, was unevenly explored. Although the lower Purus River is often mentioned in historic chronicles and colonial documents, its middle and upper reaches were only visited much later. In fact, for most of the colonial period, when Brazil was under Portuguese rule (AD 1500–1872), the present day state of Acre was part of Spain's possessions in the Americas, seldom explored by nonnatives. From the mid-eighteenth century, however, the Europeans began to travel up the Juruá and Purus Rivers in search of slaves and valuable forest products. They were looking for sarsaparilla, India-rubber, nuts, cocoa, and copaiva balm (Chandless 1866:87), often capturing young and adult natives to be sold in large Amazonian centers such as Belém and Manaus (Ferrarini 2009).

The first reports described places where natives were already acquainted with the Portuguese or Spanish language, and many tribes already made use of steel axes and knives. Contact, of course, had happened long before written reports were available. In 1818, Manuel Joaquim do Paço, the governor of Manaus, forbade the trade with the natives, a fact that gives an idea of the scope of such incursions. This sanction, however, would be lifted before the middle of the century (Castello Branco 1960). In any case, these expeditions probably did not go as far as to reach the upper portion of the Purus River basin.

In 1850, immigrants started moving to the Purus River in search of the so-called black gold: the sap of the Pará rubber tree (*Hevea brasiliensis*). The exploration of the upper courses of the Purus River became important for the economy of the upper Amazonas province, particularly after 1852, when the city of Manaus became politically independent from the province of Grão-Pará. At this point, interest in the upper Purus River derived from the need for a line of communication with Bolivia that would avoid the dangerous rapids of the Madeira River. The configuration of the Purus River headwaters was unknown at the time, and many hoped to find a waterway between the upper reaches of the Purus and the Beni Rivers. Some believed that the Purus and Madre de Dios Rivers were connected, so several possibilities were explored to provide a safe and easier way to supply Manaus with Bolivian cattle.

In 1852, when Tenreiro Aranha was sworn into office as President of Amazonas, he proposed that the routes of communication in the province be improved, demanding two expeditions, one of which should sail up the Purus River and reach the Beni River through the prairies, arrive at the Bolivian settlements, and penetrate the unexplored hinterland (Ferrarini 2009). That same year, Serafim da Silva Salgado left Barra do Rio Negro, taking four months to navigate 1,300 miles up the Purus River and finally reaching the seventh settlement of the Cucama. He brought back the important news that the Purus River was devoid of the waterfalls that had rendered the Madeira River so unsuitable for navigation. In the following years, the government of the Amazonas province created several *Diretorias de Índios* (native boards) and missions. The difference between these two types of settlements was that the *diretorias* were administered by upper-class men appointed by the government, whereas the missions were administrated by religious orders. The river was then densely populated by indigenous groups all the way up to the municipality of Pauini (Ferrarini 2009).

Between 1852 and 1860, the government of Manaus sent four expeditions up the Purus, all of them unsuccessful; none was ever able to go beyond the mouth of the Iaco River (Castello Branco 1960; Church 1877). However, they brought back valuable information on the indigenous peoples who lived along the river, as well as on navigation conditions; unfortunately, most of the information was still limited to the lower reaches of the river.

Manoel Urbano, the third explorer, was given the mission of finding a water route between the Purus and Madeira Rivers, above the Madeira River rapids; according to Chandless, he was an important source of information on the native peoples because he was acquainted with their customs and languages (Chandless 1866). Manoel Urbano was responsible for a *diretoria* located at what was then considered the upper reaches of the Purus River. He estimated that there were about 5,000 people living along the Purus River's banks. In 1861, Manoel Urbano spent 55 days traveling in canoes from the mouth of the Purus River to the Ituxi River. From there, he navigated up the Ituxi River for 100 days, passing

26 indigenous settlements of different nations along the way, such as the Ipuriña (17 villages), Juberi, Jamamadu, Canomari, and Manentenery. According to Urbano, the Manentenery were the most numerous. When he reached the mouth of the Aquiry River, he was informed that it communicated with the Madeira River, but his travels along this river proved that the information was false (Ferrarini 2009).

The geographer William Chandless ascended the Purus River in 1864 and found its headwaters in 1865, proving that it was not the source of the Madre de Diós River, and that there was no possible waterway between the upper Purus River and the Bolivian River. In a second trip (1865–66), he navigated the Aquiry River (known today as the Acre River) up to parallel 11° (more likely parallel 10°30'), from where he continued his journey on foot, trying to reach any river that would serve as a connection to the Madre de Dios basin; after a week, he had to return for provisions. He found was a number of streams running east, concluding that the Madre de Diós River met the Beni River further south (Chandless 1867). Chandless's account of the natives who lived along the Purus indicated that the river was populated along most of its course. As he was ascending the river, he found several groups: the Muras (peopling the riverbanks for 250 to 300 miles); the Puru-purus, after which the river was named; and the Hypurinás, known today as the Apurinã (along 300 miles), half of which he described as "wild" (i.e., having few contacts with the "white man" [1867:101]). After a space of 100 miles without inhabitants, he found the Manentenerys, who he described as culturally more advanced. According to Chandless, they planted cotton, made clothes, and had iron tools they had obtained from the Peruvians through indirect contact. The natives reportedly told him of a "portage from the Purus River to the Ucayali River, over which a canoe could be dragged in two days, and, when embarked on the Ucayali arrive at Sarayacu [catholic mission] in ten days" (Chandless 1867:101). After the Manentenerys, he found the Canamarys, who previous travelers had not met, and yet another group who had not had any contact with nonnatives. These observations indicate that contacts between indigenous peoples and nonnatives of Portuguese or Spanish origin happen not only through the main route provided by the Purus River, but also along smaller rivers and overland.

Given the ongoing trade in the region, Chandless (1866:88) noticed that the natives were abandoning agriculture and the production of *farinha* (manioc flour) for the extraction of wild products. Manioc flour was then imported from the province of Pará, and its price began going up.

According to Chandless, the Hypurinás (Arawak speakers), who lived between the Sepatini and Iaco Rivers, were the most numerous. They were constantly engaged in war, making use of poisoned arrows.

Chandless observed that some Hypuriná had fresh arrow wounds (Chandless 1866:96). He noted that even the people who lived close to the river always placed their huts at least a couple of miles inland, whereas others had their villages about a half-day's journey from the river. In the summer, they would move to the riverside and make temporary shelters, sometimes for collecting turtles. This is a possible explanation for the lack of archaeological sites next to the riverbanks. Moreover, the Purus River and its tributaries are known for their eroded banks and changing courses, which might have destroyed many archaeological sites through time.

Chandless also mentioned that the Hypuriná practiced secondary burial, using the bones in special ceremonies in which they praised the deeds of the dead while holding up their limbs (Chandless 1866:97). Although he does not mention urn burials, the few urn cemeteries known to archaeologists are located in the Iaco River basin (in the Sena Madureira municipality), some 200 kilometers southwest of the region mentioned by Chandless. He adds that they chewed coca leaves, keeping them in their cheeks. The landscape he describes is clearly anthropogenic: jauaru palms stretch for about 15 to 20 miles beyond the mouth of the Acre River, and bamboo becomes more common as one ascends the Purus River. In the beginning of the twentieth century, Fawcett also described plants in eastern Acre that indicate long-term human interference: "The Acre was really the Macarinarra, or 'River of Arrows,' because of the flowering bamboo from which arrows were cut. The Rapirrán, a frontier tributary of the Abuna, was the 'River of Sipos,' a vine used extensively in house construction. Another small river, the Caipera, was the 'River of Cotton,' and so on" (Fawcett 2001:83).

The territory of Acre became a Bolivian possession in 1867 via a treaty signed between Bolivia and Brazil; the Brazilian Imperial government was not really interested in that part of the territory, which was mainly inhabited by indigenous people.

In the 1870s, however, the new frontier began to receive immigrants coming from northeastern Brazil, aiming at making fortunes with a new local business: extracting rubber for the international market. Rubber extraction attracted many immigrants, especially from Ceará, a state in northeastern Brazil that has suffered greatly from prolonged droughts. Waves of migrants were then eager to explore the "black gold," as the *Hevea brasiliensis* sap was known during the rubber boom.

In the late 1870s, a commission formed by representatives of Bolivia and Brazil went to the region to establish the political borders between the two countries, and found that Brazilians had already massively occupied Acre. Moreover, waterways favored the flow of people and goods straight to the Amazon River, which was within the Brazilian territory, with no easy waterways connecting to Bolivian lands. By the end of the century, Acre was responsible for most of the Brazilian rubber exports, and Brazilians owned most of its territory, with land titles issued from Manaus. Thus, in practical terms, the territory of Acre still belonged to Brazilians, although it was a Bolivian possession.

In 1877, thousands of people invaded the upper Purus River basin, particularly the area surrounding the Acre and Iquiri Rivers. At this point, hostilities between rubber tappers and natives became more serious. Rubber tappers gradually penetrated the forest, extending their stations into indigenous territories. The natives occasionally visited these stations, threatening or killing the intruders from time to time. The natives who resisted and did not join the missions or Native boards were regarded savages. It was not uncommon for rubber tappers to capture native women for wives. The encroaching whites would frequently promote raids to enslave the native population for the rubber industry, a situation that 40 years later would result in the near-extinction of tens of thousands of natives (Rusby 1933). With the high prices obtained for the rubber, the region began to develop very quickly. In 1881, the Purus River yielded many items for exploration, such as rubber, Brazil nuts, copaiva oil, dry fish (*pirarucu*), sarsaparilla, and cumaru (*Dipteryx odorata*).

In 1887, the government of the province of Manaus ordered Colonel Antonio R. P. Labre to seek an overland corridor between the indigenous rubber settlements along the Madre de Dios River and the nearest navigable point on the Acre River, a tributary of the upper Purus River, in hopes of building first a road, and later a railroad that would connect Bolivia to the city of Manaus through the Purus River (Labre 1888, 1889). In this mission, Labre ascended the Madeira, the Beni, and the Madre de Díos Rivers. From the Maravilha port, Labre crossed overland until he reached the Acre River. In doing so, he went through several Araona villages, some of them abandoned, perhaps because of raids, missionization, and rubber extraction activities. On the sixth day of the journey, he arrived at the Araona village of Mamuyeçada, which was inhabited by about 200 people. According to Labre, they had an organized government, religion, and temples. Their gods were represented by carefully polished wooden idols, which, unexpectedly, did not display any recognizable human form, but were instead geometrical. The father of these gods was represented by a 16-inch-tall elliptical-shaped wooden object called Epymará; these idols were likely kept in temples inside the forest (Métraux 1942).

Labre did not describe the temples, but his account is especially informative, for it yields a clue to the possible meanings for dozens of geometrical earthworks built by indigenous societies in the vicinities of the Araona villages—the enclosures known as geoglyphs. They were built at carefully chosen places in the higher terrains, and display few vestiges that would not attest their use for habitation purposes, appearing instead to be religious and ceremonial centers.

The Araonas and other Takana speakers were densely populated along the Madre de Díos River, which they called the Mayu-Tata. According to Father Fidel Codinach, a Franciscan friar, "Mayu" is a prefix used to designate a large river; thus, Mayu-Tata in Takana means the Great Father River. The Qhichua (Quechua) speakers called it Amaru-Mayu, or the Serpent River (Church 1877:96). In addition, the Takana talked about another large river to the north, a short distance from the Mayu-Tata, which was likely the Aquiry (Acre) River. According to the friar's manuscript (Church 1877), the Araonas, a large division of the Takana, lived between the Madre de Díos and this other river, in a territory that today is occupied by Brazil and Bolivia.

In 1897, the borders between Brazil, Peru, and Bolivia were still a matter of debate, so another commission was created to establish the limits. At the end of the century, the upper Purus was nearly completely occupied by Brazilian rubber tappers. However, between 1898 and 1903, Bolivia established a post in Xapuri to take possession of the territories ensured by the 1867 Ayacucho treaty. In 1899, they officially planted the Bolivian flag at the Acre River, claiming its possession, and thus heating up the dispute for the territory. This feud involved an armed conflict (known in Brazil as the Acrean Revolution), a peace treaty (signed in 1903), and a reimbursement made to Bolivia, which more than paid off, given the amazingly high revenue Brazil would obtain from the rubber business in the few years that followed.

By 1912, the growing Amazonian economy, largely based on the extensive exploitation of wild rubber, faced a fatal challenge: competition with the rapidly expanding rubber plantations in Asia, which offered more competitive prices in the global market. The decentralized production system in Acre and other parts of Amazonia, which involved high risks and high transaction costs, could not compete with the labor surplus that enabled lower-priced products coming from Asia. With no investments in any other sector, such as agriculture or manufacturing (Barham and Coones 1996), the region saw a capital flight, and the population was forced to turn to a subsistence economy.

Despite its disadvantages in the global market, rubber continued to drive the region's economy during most of the twentieth century. The rubber flowed through patrons and traders established in the cities of Belém and Manaus whereas the indigenous peoples were systematically pushed out of their lands in bloody encounters between natives and non-native rubber tappers.

Because the region lacked any other economic appeal, in the 1970s, the military government decided to populate it with immigrants drawn from oversettled farming communities in the south as well as from the poorest rural areas in the northeast, a policy that was extended to other regions in the Amazon.

Upon arriving in Acre, these hopeful families had no other choice but to cut down the forest to plant their crops and raise livestock. The sale of farm crops was not rewarding, however, because markets were too far away and roads were precarious and therefore inappropriate for transportation. As a consequence, in the 1980s and 1990s, agricultural fields were turned into pasture for cattle, which was considered a more viable economic strategy. Since cattle raising demands larger fields, deforestation rampaged across the region.

Ironically, deforestation in this case turned out to be a benefit to archaeologists. As farmers and ranchers cut down trees, some extraordinary earthen structures became visible to their curious eyes. Some of the landowners would not even bother to guess the form of the figure drawn on the ground, whereas others would walk along the ditches only to find out that they represented closed shapes, like rectangles, squares, and circles. Ranchers and farmers were convinced that there was only one plausible explanation for those structures, and it was not a natural one: they thought they were defensive ditches built during the Acrean Revolution, the war against the Bolivians.

## The Indigenous Population and Possible Meanings for the Geoglyphs

The available historical information is very elusive when it comes to the possible meanings of the geometric sites. During the colonial period in Bolivia, the ring ditches were believed to have served a defensive purpose against the feared Guarayos (Eder 1985); nobody has witnessed people living on geoglyphs in Acre, with the only possible exception of Chandless, who mentioned a ditch in an indigenous village serving the possible function of protecting a house used to store feast materials.

From the reports of explorers who traveled up the Purus, Acre, and Iquiri Rivers, we can infer that this region was occupied primarily by Arawak populations. In the beginning of the twentieth century, Percy Fawcett journeyed through Acre, covering the territory adjacent to the Bolivian border. His account of the area encompassing the vicinities of the Acre River, where the municipality of Capixaba is located today, and where more than 30 enclosure sites are located, is highly informative:

> We stayed at a place called Campo Central in order to trace the sources of certain rivers and find their position. While doing this, we came upon enormous circular grass clearings, a mile and more in diameter, the site a few years earlier of large villages of the Apurina Indians. A few of these Indians still lived at another place called Gavion, and some others, lucky enough to escape slaving expeditions, fled northwards. ... They

buried their dead in a sitting posture, and everywhere in the clearings were graves (Fawcett 2001:81).

On the other hand, the fact that many sites are located next to the Bolivian border, in a region that remains largely unexplored and where Labre found the Araona in the nineteenth century who allegedly adored "geometric gods," indicates a possible religious meaning for the enclosures. According to missionaries, the Araona and other Takana speakers lived between the Madre de Diós and Acre Rivers for a long time, so it is not unlikely that they were the ones responsible for building the enclosures in that area.

In the beginning of the twentieth century, Takana-speaking groups inhabited a "contiguous territory that included the upper courses of the Orton, Abuna (which divides Brazil and Bolivia), and Acre rivers; the Madre de Diós River between longitudes 67° and 68°35', and its tributaries; the Tambopata and Heath rivers; the Beni River between latitudes 12° and 15° S, and its tributaries, especially the Madidi and Tuichi rivers" (Métraux 1942:30). After many changes caused by colonization (e.g., missions, enslavement, and wars), the remaining Takana, who originally consisted of several intermixed groups that spoke dialects deriving from the same mother language, ended up borrowing words from other groups, which confused linguists. As the available linguistic material, which allows for the classification of all of the Takana groups into a same family, was reduced, Métraux considered that there were many resemblances between the Takanan and Panoan dialects, and that the Takana had also incorporated a large percentage of Arawakan words. This is why "Créqui-Montfort and Rivet (1921–22:147) were predisposed to consider the Takana as an Arawak dialect which had been strongly influenced by Panoan languages" (Métraux 1942:30). The official language spoken in the missions was Quechua, which helped make circumstances even more confusing.

The first contact with the Takana is supposed to have occurred in the sixteenth century, when expeditions left Peru for the Mojos, likely penetrating the Takana territory on the way (Métraux 1942). The 1567 Maldonado expedition established contact with the Takana, and Fray Gregorio Bolívar also described many Takana groups in 1621. In 1678, Franciscan missionaries visited Takana; after that, many expeditions were organized to establish missions among the Araona and Tomorona (Armentía 1887; Métraux 1942). According to Maurtua (1906), during the 1678 visit, some of the missionaries destroyed the idols kept in a temple, replacing them with the image of a saint (Métraux 1947).

The Takana cultivated several different crops in small plots; the Tiatinagua chose tracts covered with bamboo, which were easy to cut using stone axes. They would also plant bananas and plantains along the rivers so they could harvest them during their hunting and fishing seasons. The Tiatinagua cultivated yucca, maize, sweet potatoes, malanga, gourds, tobacco, cotton, and sugarcane; besides these, the Araona planted papayas, tubers, and coca. All of them collected wild food, especially palm fruits and Brazil nuts. The Takana took advantage of fishing in shallow waters, placing rectangular enclosures across streams.

Although the historic Araona lived mainly along the Madre de Diós River, sources make it clear that they also knew the Acre and Purus Rivers. In the last quarter of the nineteenth century, there was a lot of population migration in the region due to rubber exploitation and quickly spreading diseases, which led natives to flee from all contact with the nonnatives. For instance, many Araonas migrated to the Abunã River in 1885, trying to escape a measles epidemic (Armentía 1887).

The region where the geoglyphs are located seems to be enclosed between Arawak and Takana territories, so it is feasible that both groups were responsible for them (Figure 5.11). An interesting hypothesis is that the Takana would have built the circular and elliptical ditches located in the south, whereas the Arawak would have been responsible for building the square and rectangular sites situated in the north.

Pirjo Virtanen, a Finnish anthropologist who has conducted research in Acre with the Apurinã, a group that belongs to the Arawak linguistic family, believes that the consumption of ayahuasca in traditional

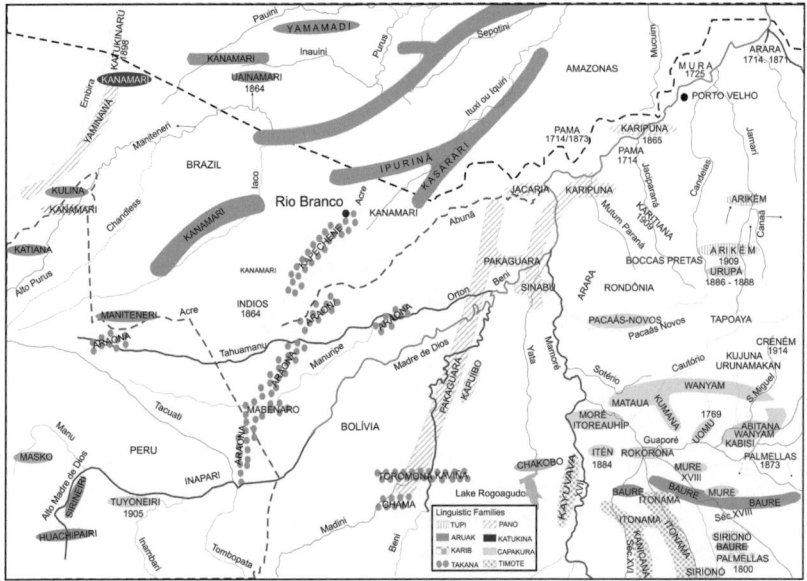

Figure 5.11. Map of the western Amazonia showing the locations of indigenous groups. Based on Nimuendajú (2002).

religious ceremonies could have had some relation to the geoglyphs, as the drug provokes visions of geometric figures usually interpreted as supernatural beings. It is reasonable to infer that for those peoples, their gods would assume geometric forms that had to be marked on the earth as their temples. Virtanen has also worked with the Manchineri (Arawak), who have no memory of the ditched enclosures, but have furnished some interesting information regarding their cosmologies. The Manchineri used to live along the Purus River and its tributaries, the Aracá, Caspaha, and Yaco Rivers (e.g., Chandless 1866; Steward and Métraux 1948). Present-day Manchineri, though, say that their ancestors lived along smaller tributaries as a way of protecting themselves from hostile groups, and later on, in order to hide from the eyes of nonnatives, who enslaved the natives to work in the rubber extraction industry. To the Manchineri, self-protection has been a crucial factor in choosing where to live. Palms abound in the region, and the Manchineri relate palm spirits to ayahuasca visions. The palm spirits appear and the designs "arrive" when nonhuman knowledge becomes available. Furthermore, the same ayahuasca vision containing iconography that is similar to the earthwork structures, as mentioned earlier, appears also in connection with *Pwernokatu,* a Manchineri ancestor. The design is visible on his clothes whenever he is represented in drawings or described after ayahuasca visions (Virtanen 2010).

## Chapter 6

## CONCLUSION

## A Landscape Perspective of Amazonian Pre-Columbian History

FOR SEVERAL DECADES, the notion that Amazonia was a land without opportunities for the development of complex societies and cultures was the prevailing view of anthropological research in the region. In Brazil, those scholars who held with this viewpoint left little room for dissonant voices in their own territory. Students with a different perspective and alternative research designs were not welcome to excavate sites or have access to museum collections in Brazil (Roosevelt 1991; Schaan 2004). Data produced by scholars who were not aligned with mainstream archaeology were eventually discarded or modified to maintain the status quo of a number of outdated archaeological postulates (Roosevelt 1995; Schaan 2001b). Cultural ecology was the dominant theoretical framework (Meggers 1954; Steward 1948, 1949), a perspective that emphasizes systems theory, the ecosystem, and adaptation, leaving no room for human agency and cultural change (Balée 2006).

Scholars in several related fields of inquiry began to criticize these views early on. Departing from his fieldwork experience among the Kuikuru, Robert Carneiro (1960, 1961) contributed with important data on indigenous subsistence systems, insisting that nomadism and low population levels were not caused by local ecological conditions, since the Kuikuru had managed to live in the same areas for almost a century, relying on manioc cultivation and fishing. Carneiro's research became important because the region occupied by the Kuikuru was far from the Amazon floodplain and provided data on peoples who lived in an environment that was considered inadequate for cultural development. In the mid-twentieth century, important distinctions were drawn between *várzea* (the Amazon floodplain) and *terra firme* (nonflooded areas or high ground between rivers), anticipating a concept that involved the variability of Amazonian landscapes, a concept that would only be fully understood decades later. Lathrap (1970b) then proposed that given its high productivity in terms of aquatic fauna and fertile land (because of annually renewed rich sediments), the Amazon River floodplain would be a population magnet and a place for experimentation and the development of advanced techniques as well as cultural ideas that would eventually be exported to the whole basin through the Amazon's fluvial network. Carneiro called Lathrap's interpretation a "cardiac model" in which the *várzea* would be the heart, and the Amazon tributaries the arteries from where people would spread out to progressively occupy the hinterland, maintaining complex exchange networks, social interactions, warfare, and alliances that promoted the flow of goods, peoples, and ideas, expanding an advanced way of life which Lathrap called "the tropical forest culture."

Steward (1948) and Meggers (1971) also provided a novel assessment of the supposedly superior role of the *várzea* in the Amazonian cultural development, although they interpreted the distinctions between these two broadly described ecosystems as related to their different scope (the *várzea* offering more opportunities for demographic increase), rather than to their general role as a hindrance to the indigenous population's presumable aim of overcoming nature's constraints. According to Meggers and Steward, some of these constraints included poor soils, the lack of steady protein sources, and cyclical climatic changes, variables human beings could not control. As a consequence, Amazonian populations would have been highly mobile, changing their cultivation plots regularly because of soil exhaustion. Without possibility of agricultural intensification, population levels would remain low, and complex sociopolitical institutions would not develop (Meggers 1995).

Theory wars had to be supported by field data. On the cultural ecologists' side, data were provided by physical geography, ecology, and a vicious system of archaeological settlement analysis based on artifact seriation, using Ford's methodology. The method consisted of identifying sites, collecting ceramics from the surface, or test excavations using arbitrary levels. Potsherds were then classified into types and used for building a chronology without regard for cultural features that could help in providing context. This methodology was used to back up the underlying theory (Meggers and Evans 1970): stratigraphy was considered nonexistent (Evans and Meggers 1965); landscape features were deemed irrelevant; and archaeological evidence of social complexity (social rank, monumental earthworks, and craft specialization) was interpreted as traits brought by peoples who came from the more complex Andean civilizations. At that time, the most significant data to contradict the "standard model" (Viveiros de Castro 1996a) came from the field of historical geography.

Building on previous studies by Nordenskiöld (1930) and Sauer (1969), Denevan (1963) took on the task of looking at the Amazon from a landscape perspective. His research on the Bolivian Amazon (see Chapter 5) represents one of the most enduring and important contributions to Amazonian archaeology. Denevan observed that human societies in the Bolivian lowlands had altered the landscape to such considerable proportions that this would have overcome any limitations that natural conditions could have had on their subsistence strategies and ultimately in their lives. According to Denevan, the practice of raised field agriculture and causeways above annual flood lines was clear evidence that the native inhabitants were creating physical conditions for growing crops to feed large populations and dominating wetlands. More recently, Erickson (2006), who followed in Denevan's footsteps, has characterized human-transformed environments as "domesticated landscapes." According to Erickson, the concept of domestication applies to "humanized" landscapes, or a landscape modified according to human needs and beliefs of what a landscape should be (Erickson 2006): "Domestication of landscape implies all intentional and non-intentional practices and activities of humans that transform the environment into a productive landscape for humans and other species. Domesticated landscapes are the result of careful resource creation and management with implications for the diversity, distribution, and availability of species" (Erickson 2008:158). Heckenberger (personal communication, 2008) observes that the word "domestic" derives from the latin "domus" (i.e., house or residence). Domestication as applied to landscapes would be transforming the landscape into home, a place centered on people and their demands.

The idea that Amazonian societies had overcome environmental limitations was crucial for the debacle of the unitarian view of sociocultural development in the lowlands. A better understanding of the Amazon as a region of multiple contrasts emerged among biologists, geographers, and environmental anthropologists (Denevan 1976; Morán 1995; Prance and Lovejoy 1985). The two alleged major ecosystems—*várzea* and *terra firme*—were shown to simplify more diverse environments with distinctive geological histories and vegetation patterns as well as varying degrees of fragility to exploitation and different histories of human management (Morán 1995).

Today, most students of the Amazon recognize that indigenous societies have left enduring marks of their presence in the tropical lowlands. Some of these transformations have permanently affected local topography, remaining visible to this day, as archaeological sites are being reused and altered by recent populations. Some examples of this are the raised fields in the lowlands of Bolivia, Venezuela, and the Guianas; mound platforms and other earthworks in Marajó, Ecuador, and Bolivia; and the ubiquitous ADE, used for cultivation throughout the basin. Other transformations in vegetation patterns are subtle, and their time depth can only be evaluated by environmental history.

Ethnographic, botanical, geographical, and ethnoarchaeological research have demonstrated that forest management practices have contributed to the present biodiversity. Hunter-gatherers manage plants along their paths, carrying seeds that are discarded along their hunting trips (Politis 2001; Posey 1985). This practice thus becomes a plant dispersal factor. Plant management, in its diverse forms, has preceded the development of crop domestication, since agricultural techniques are believed to have evolved from the previous knowledge of local plants' life cycles and their use by humans. Agricultural societies have also affected forests, as totally new forest types are found in abandoned plots that feature the dominance of certain species, especially those useful to humans, such as palms and fruit trees. Over the years, these disturbed, anthropogenic forests attract new or returning populations; as a result, their current constitution is the result of centuries of human management. Balée (1989:14–15) has proposed that at least 11.8 percent of Brazilian Amazon's *terra firme* is composed of anthropogenic forest, a figure that has not been disputed. On Marajó Island, mounds that were once inhabited are now anthropogenic forest islands, containing important economic species such as açaí (*Euterpe oleracea*), bacuri (*Platonia insignis*), Brazil nut (*Bertholletia excels*), buriti (*Mauritia flexuosa*), cacao (*Theobroma cacao L.*), embaúba (*Cecropia peltata*), inajá (*Maximiliana maripa aublet drude*), mango (*Mangifera indica*), pupunha (*Bactris gasipaes*), silk cotton tree (*Ceiba pentandra*), taperebá (*Spondias lutea*), tucumã (*Astrocaryum vulgare*), and urucum (*Bixa orellana*), among others. In the savannas, mounds and their refuge forests are little oases in a wetland landscape dominated by grass and shrubs.

Humans have altered Amazonian landscapes in different ways, either intentionally or as a side effect of their cultural practices. Earthworks are the result of the intentional removal of soil from the ground to produce ditches, ponds, sunken plazas, roads, and canals; the resulting earth is then used to create elevated structures such as raised fields, platform mounds, walls, fish weirs, and raised causeways, involving the coordination of dozens or even hundreds of people. This conscious task was directed toward significantly modifying the surroundings, with enduring consequences for local ecology. In most cases, earthworks imply a monumental construction effort performed for economic reasons, requiring maintenance for decades and even centuries. Archaeologists have found enormous earthworks that, centuries after abandonment, were covered by forest, giving the impression of a pristine landscape (Erickson 2010; Erickson and Balée 2006). In these cases, a complete species turnover is a consequence of what Balée (2010) has called "primary landscape transformation."

Our accumulated archaeological knowledge of landscape modification across the Amazon basin indicates that several hundred kilometers of continuously modified landscapes were subject to this complete species turnover. Balée's early estimation of 11.8 percent, however, is conservative, because archaeological data were scanty at that time (Balée 1989). For example, most of the state of Acre has been transformed by human occupations that preceded the Europeans' arrival. The presence of geoglyphs over an area of about 30,000 square kilometers, coupled with the accounts of early explorers traveling along the Juruá and Purus Rivers (Chandless 1866; Church 1877; Fawcett 2001), who reported extensive patches of plant species characteristic of anthropogenic forests, indicate that at least a third of the state (50,000 square kilometers) has been altered by humans.

On Marajó Island, the 20,000-square-kilometer savannas have been modified by pre-Columbian populations through the construction of dams, mounds, ponds, and causeways (Schaan 2004, 2008). At Llanos de Mojos, in Bolivia, the 90,000-square-kilometer region of savannas and forests shows evidence of human management (with mounds, raised fields, canals, and causeways). In addition to earthworks, these areas have probably been maintained as open landscapes due to systematic burning, presumably a pre-Columbian practice (Denevan 1966; Erickson 2006). So, as we enlarge our body of data, including information from both archaeology and ethnobotany, we will likely find that at least 50 percent of greater Amazonia has been significantly altered along the history of its human occupation, since 11,200 years BP.

Landscape management was accompanied by important inventions and cultural developments, resulting in significant transformations within the Amazonian social systems. The initial domestication of crops is estimated to have occurred during the same time period as the oldest pottery found along the lower Amazon (7,000 years BP). The emergence of unique art styles, such as the Saladoid-Barrancoid Horizon, which influenced several Formative ceramic styles along the central and lower Amazon, penetrated the Amazon basin through the Orinoco River, coming from Venezuela. After AD 400, the Polychrome Horizon pottery makes its appearance, along with the proliferation of earthworks throughout the basin. New regional social systems were advanced by peoples who combined landscape management and special connections to the supernatural world. Art styles, especially those used to decorate funerary and ceremonial ceramic vessels, indicate the existence of religious cults for the ideological justification of new rulers and their factions.

Amazonian polities, which were fully developed a few hundred years before European contact, should be studied in view of their unique development vis-à-vis other complex societies around the world. Some authors have avoided applying the term "hierarchy" to those social systems, which seem to have been more egalitarian than other chiefdoms or state societies studied elsewhere. In some cases, the archaeological absence of a central place to which other satellite settlements would be subject, along with conspicuous signs of social rank, are taken as indications of heterarchical social systems (Heckenberger 2005; Pärssinen et al. 2003; Roosevelt 1999). Crumley (1987, 1995) uses the term "heterarchy" to characterize multiple ways through which units of settlements and societies might be ranked to one another, considering temporal and spatial scales. Clearly, when it comes to settlements and social systems, both hierarchy and heterarchy exist to a different extent, depending on the scale of a given social organization. It seems that some Amazonian complex polities exhibited several local hierarchical social systems under a heterarchical umbrella of regional organization. Depending on particular historical conjunctures, a more centralized structure would probably be necessary to overcome a common threat, such as needing to face a common enemy (Carneiro 1998).

Complex societies in Amazonia did not represent the top of the ladder in a continuous and unavoidable evolutionary process; instead, they emerged in specific places at specific times, according to local conditions. These conditions were occasionally debated, as many scholars have weighed in on this matter (Carneiro 1970, 1998; Roosevelt 1980; Schaan 2004). What I wish to emphasize, however, is that complex societies in Amazonia coexisted with several more numerous and distinct social formations, with which they built relations that involved alliances, trade, warfare, symbiosis, emulation, and avoidance. Acknowledging this configuration places us in a position of rejecting evolutionary, linear models in the analysis of complexity in the Amazon, but this should not exempt us from researching those important themes, which might illuminate our understanding of sociopolitical and cultural change.

Several scholars have used ethnoarchaeology and ethnographic analogy as resources for understanding and explaining pre-Columbian social systems in Amazonia, based on either cultural continuity or the survival of a long-term Amazonian ethos (Gomes 2007; Heckenberger 2001a, 2005). Heckenberger has studied the long-term history of Arawak occupation of the upper Xingu River, arguing for a cultural continuity in the last 800 years, despite the huge demographic collapse and population movements that happened in the colonial era. Cultural continuity is represented by the regional sociopolitical organization of hierarchical polities expressed archaeologically in particular site layouts that have survived to the present (Heckenberger 2010). In the upper Xingu, ethnography and oral history have provided detailed descriptions of male ceremonies that imprint on the body the marks of social inequality (Heckenberger 1999, 2005). Cultural practices not present in the archaeological record are then assumed to have a deep history, which began with the arrival of Arawak in the region by AD 1250. Gomes (2007) believes that scholars have overemphasized the historical discontinuity between pre- and post-contact Amazonian populations, and that ethnology has much to contribute to understand the sociocosmology of these societies. Gomes feels that the study of pottery iconography, which would mirror "social morphology," does not include clear signs of inequality (Gomes 2007:214).

The fact that regional, complex societies no longer existed as they did in the fifteenth century and were not recorded by ethnographers challenges the validity of ethnographic projection, as careful as that might be. Amazonian scholars are yet to produce a theoretical framework that would set limits to ethnographic analogy as well as define what terms can be compared and under which analogical argument. Ethnoarchaeology certainly holds the key to such an effort, if combined with a serious approach to past societies that would balance structure and history (Politis 2007; Silva 2010).

## Sacred Geographies of Ancient Amazonia

The pre-Columbian Amazonian societies and their achievements can only be reconstructed through archaeology, since ethnohistorical sources are scanty and biased; their use is possible only when confronted with the archaeological record. Archaeology is multidisciplinary in nature, as it deals with the material remains of human societies; these remains have been affected by several physical and chemical processes. So, to accomplish its enormous task, archaeology needs the help of disciplines such as paleobotany, paleoenvironment, geoarchaeology, pedology, biological anthropology, linguistics, and zooarchaeology, to name a few, which have long been archaeology's allies and are of critical importance in Amazonian research. The endeavor of reconstructing the history of indigenous people in the Amazon basin also needs a long-term, dialectic perspective on the coupled relations between human societies and tropical environments, a perspective that includes agency and history. This perspective is provided by historical ecology (Balée 1998, 2006; Crumley 1994).

Historical ecology and archaeology are important for present populations for many reasons. Telling the history of the changing Amazonian landscapes, of which humans are the critical agent, means rescuing Amazonian peoples from the invisibility to which cultural ecology and other disciplines emphasizing adaptation as well as a disconnect between nature and culture have relegated them. To cultural ecologists, indigenous peoples have a history of symbiosis and adaptation to nature. This implies an eternal state of being, without history. A fair reconstruction of the indigenous past provides them with history, promoting the valorization of indigenous peoples, with important consequences relating to the acceptance of their way of life, social identities, and territory rights by society at large.

Present peoples' memories and knowledge of the areas they occupy are seldom heard or used in environmental impact studies (Beltrão 2008; Crumley 1994), including archaeological research within CRM projects. Present techniques of landscape and wild resource management mirror traditional indigenous practices in many settings, yet archaeologists rarely ask the populations living in and around archaeological sites about their knowledge on the history of meandering rivers, tide regimes, flood management, soils, choice of house location, selective use of woods and palms, and local histories of forest succession. Present practices of landscape use are part of an enduring history of peoples and places that need to be placed in a same continuum. Archaeologists working in Amazonia need training in four-field anthropology, a discipline tradition that has been contested in the United States (where it was born) for many reasons, such as the diversification of interests among anthropologists (Hodder 2005; Segal and Yanagisako 2005), but which remains important for the study of Amazonian populations.

In response both to an international trend toward more inclusive research projects as well as to demands by the populations being studied, a growing number of archaeological projects have included native and traditional peoples in their research teams, establishing greater levels of dialogue and participation in their projects.

Archaeology among the Kuikuru (upper Xingu), the Palikur (northeastern Amapá), the Asuniri (Xingu), and the Kayabi (Mato Grosso and Pará) has involved surveys aimed at finding prehistoric villages; the indigenous people have indicated the location of these sites based on the occurrence of archaeological materials and anthropogenic soils as well as their collective memories (Green et al. 2003; Heckenberger et al. 2003; Silva 2002, 2009). Coupled with archaeological evidence, oral history has been used by archaeologists to propose a long-lasting occupation history of the same territories by the same ethnic groups (Heckenberger 2001b, 2005). Oral history and indigenous interpretation of archaeological artifacts and places have also offered alternative indigenous narratives that contrast with the perspective of Western science (Silva 2002). The Asurini, a Tupian population living along the Xingu River in the state of Pará, interpret smooth, round concavities on river rocks (called lithic tool workshops by archaeologists) as marks left by a mythical hero's feet and buttocks, while ceramic sherds are believed to be *Anumai*'s broken pots. In a mythical story, she breaks all her pots on *Tapijawara*'s head to force him to go back to the underworld (Silva 2002:177–180). Archaeological stone tools found in their territory are used by the Asurini today for many daily tasks, while new ones are produced using the same old technology, a sign that they have preserved cultural memory, and have not totally abandoned old industries, as did other groups (Silva 2002:181–182).

Indigenous participation in archaeological projects frequently has clear political agendas, either to reinforce oral stories and ethnic identity, or to advance claims for historical territories (Green et al. 2003;

Silva 2009), especially when limits are being questioned. The practice of archaeology in indigenous lands can benefit from indigenous information and help, but a dialogic relationship is necessary to fulfill both the researcher's pursuit of knowledge and the natives' need for recognition. Moreover, the acceptance of the possibility of multiple views of the past is an important step in providing social room for multiculturalism in a democratic society.

Amazonian archaeology has not yet investigated territories occupied by runaway slaves during the seventeenth and nineteenth centuries, who, like many Indian groups, escaped from the Europeans, building *quilombos* (settlements of African fugitive slaves) at river headwaters and other hidden places. Trade and intermarriage between *quilombolas* and Indians are both reported in historical documents, oral history, and through DNA research (Ândrea Santos, personal communication, 2010). Communities of *quilombola* descendants in Amazonia practice subsistence techniques, such as manioc cultivation, fishing, hunting, and the collection of Brazil nuts, that parallel indigenous practices and have partially survived to these days (O'Dwyer 2006). It is not uncommon to find African descendants living on top of indigenous archaeological sites. Their right to communal territories was granted by the Brazilian Federal Constitution in 1988; many have already obtained land rights, and several others are in the process of doing so. At the lower Amazon River, there are about 60 *quilombola* communities in the municipalities of Oriximiná, Óbidos, Santarém, Monte Alegre, and Alenquer. At the Trombetas River, the Boa Vista community, the first to obtain a land title in 1995, occupies an area situated 70 kilometers upriver from Oriximiná. Before an anthropological report was produced (Acevedo-Marín and Castro 1998), that population was characterized as "riverine," without concern for their ethnic identity. They were encroached upon, having lost a good part of their territory to a National Forest Reservation, which they used to exploit extensively, especially collecting Brazil nuts for sale. Their traditional subsistence practices are seen by IBAMA (which is in charge of natural reserves and protected areas) officials as having a destructive impact on biodiversity (O'Dwyer 2006).

On top of a high bluff in the Boa Vista community area, a huge ADE site was excavated by Peter and Klaus Hilbert (1980) and Vera Guapindaia (2009). They found that the site was occupied from 65 BC to AD 205 by peoples associated with the Incised Rim Tradition (Pocó phase), and from AD 1200 to AD 1400 by populations related to the historical Konduri (see Chapter 4). Sometime between the nineteenth and twentieth centuries, a group of African slave descendants who had lived along the Trombetas Falls arrived at that part of the river, establishing new settlements.

During the 1970s, the mining company Mineração Rio do Norte was established in the area to extract bauxite, a rock from which aluminum is obtained. In addition, a CRM project related to the Impact Mitigation Program, run by the Goeldi Museum from the city of Belém, planned activities for the *quilombola* population. Through this project, a group of women and men were trained to manufacture pottery emulating archaeological ceramics (Konduri phase). The objective was to rescue indigenous practices (Schaan 2007).

Treated as riverine peoples, the *quilombolas* have maintained a strong sense of ethnic identity within a context of fluid cultural boundaries. To them, the notion of an ancestral territory from which traditional practices cannot be disentangled is a heritage that keeps them connected as a group. Although their ancestors inhabited the headwaters and came to the settlement of Boa Vista only in the last century, this is considered their ancestral land, founded by a great-grandfather whose magical powers and accomplishments are constantly reinforced whenever they get together for celebrations or any other special occasions in which families are reunited.

As I have argued throughout this volume, the relations between social groups and places have produced landscapes that are not only impregnated with the history of their physical transformation, but have also been interpreted by traditional societies as cultural products in a deeper, symbolic way. Fernando Santos-Granero (2004) offers a sensitive analysis of these relations in his assessment of the historical meanings the Yanesh people of the Peruvian lowlands attribute to their landscape. Rocks, salt mines, vegetation, and rivers are all connected to their ancestral history. According to their account, for instance, cultural heroes were transformed into rocks, and the grassy landscape was formed as the result of a fire caused by a fight between deities. Thus, the morphology of the landscape is seen as a product of divine interference. Therefore, not only the landscapes that are shaped during one's lifecycle are embedded with a sense of place, but the whole territory where one lives and circulates is full of historically, culturally relevant landmarks (Santos-Granero 2004).

In the Brazilian Amazon, the construction of hydroelectric dams, backed by environmental impact studies exhibiting serious flaws, has jeopardized territories occupied by indigenous and other traditional communities as well as their cultural landmarks. In several instances, anthropologists and archaeologists hired by CRM companies, most of whom have no serious anthropological training in the specifics of Amazonia, have not considered local meanings and cultural history in their reports, privileging material evidence, and disregarding the fact that in Amazonia, nature and culture are not separate entities.

Ethnologists have taught us that many Amazonian indigenous groups incorporate animals and inanimate objects into the same social realm where humans belong. In this sense, culture is a universal character, and biology, or nature, the particular way groups of animals or humans present themselves (Viveiros de Castro 1996b). Viveiros de Castro (1996b:116–117) has argued for an "indigenous theory according to which the way humans see animals and other subjectivities populating the universe—deities, spirits, the deceased, inhabitants of other cosmic levels, weather phenomena, vegetables, and sometimes even objects and artifacts—is profoundly different from the way these beings see humans and themselves." Therefore, animals would see themselves as humans, assuming a human form while in their villages, so that the appearance of each animal species is only a dress that covers an underlying human form.

To the Achuar of the upper Amazon basin, natural and supernatural realms are part of a same continuum, where nonhuman beings share the same ontology as humans: "humans and most plants, animals, and meteors are persons (*aents*) with a soul (*wakan*) and an individual life... if nature's beings are anthropomorphized, it is because they are assumed to have the same faculties as humans, even though their appearance differs" (Descola 1994 [1986]:93).

Interestingly, the idea that humans may have been animals and vice versa, or the possibility that these two might transform into one another, has survived to this day among non-Indian natives. In the 1970s, a Jesuit father recorded non-Indian natives in Marajó Island stating that monkeys were once human: something he found very amusing, since according to Western science, humans evolved from apes instead (Gallo 1996).

Everywhere in the Amazon region, a popular tale in which a dolphin transforms itself into a handsome man and flirts with young women is taken seriously; many people affirm having seen him. Although this tale is used as an excuse for unwelcome pregnancies as well as to save men from taking responsibility, especially when he is a rich, powerful rancher, and the girl is the daughter of an employee (Vianna 1998), the fact that people believe it in the twentieth century indicates that indigenous cosmologies are still present in contemporary thought systems.

These examples suggest that a cosmogony that socializes nature has a long history among Amazonian peoples. The transformation of the shaman into a jaguar or other beings is a recurrent theme in indigenous narratives both in the present and the past (Reichel-Dolmatoff 1971). Our only gateway to pre-Columbian indigenous cosmologies is the study of artifact iconography. In the iconography of archaeological ceramics and lithic tools, there are profuse depictions of hybrid (human/animal) creatures, or representations of animals and humans placed side by side or on top of each other, which mimic metamorphosis (Gomes 2010; McEwan 2001; Porro 2010; Roosevelt 1991; Schaan 2009a).

As I argued in Chapter 3, the iconography of funerary and ceremonial pottery on Marajó Island played an important role in legitimizing the status quo in a ranked society. More importantly, it helped ensure group membership as well as build strong bonds with the supernatural world that would guarantee fish abundance. "Cosmologies and the ceremonial centers that they generate have an absolutely essential role in dissipating the disruptive tensions that human society generates. A loss of faith in the ceremonial life of the community is as fatal to the polity as a loss of sufficient calories to feed its people" (Lathrap 1985:242). It is possible that in Marajó's case, an unexpected climate change affecting subsistence systems would have triggered a loss of faith in both the ceremonial life and the leaders of the community, leading to the collapse of its sociopolitical institutions.

Authority, then, was not based on coercion, but embedded in a delicate balance between supernatural and natural powers, within societies that did not perceive these two ontological orders as being separate. According to Heckenberger (2005:333), Xinguano chiefs have exclusive access to and control over some means of social and symbolic reproduction through ritual; this source of power could eventually be used to advance other interests. Rituals and ceremonies are performed in plazas or ceremonial centers that are part of the landscape, the same landscape that is being historically built—or domesticated, in Erickson's terms (2006)—and socialized through mythical narratives.

According to Santos Granero, the landscape is sacred to the southern Arawaks (as he calls the Yanesha of the Peruvian lowlands), who see it as an animated entity that they reverence through their everyday practices of respect to its rules. This notion of sacredness might be extended to all built landscapes in Amazonia, since indigenous societies have imbued them with culture and history, transforming them into the lands of their ancestors. They definitely have built a sacred geography that Westerners have mistaken for a pristine wilderness.

## How Much of the Future Lies in the Past?

Today's Brazilian Amazonia populations comprise a mosaic of peoples ranging from natives to immigrants from all over the country. In AD 1500, Amazonia's native population was between 5 and 6 million people (Denevan 2003) and spoke about 700 different languages (Rodrigues 2001). These figures were drastically reduced during the first two centuries of conquest, with a catastrophic population decline and the consequent loss of linguistic and cultural diversity. It is estimated that 80 percent of the Amerindians died from diseases caused by Old World bacteria (especially smallpox and measles), battle fights, and slavery. Population losses can also be attributed to illnesses caused by a lack of provisions due to foreign encroachment as well as the disruption of indigenous economic and governmental systems, especially those based on hydraulic works and kin-based leadership (Wolf 1997:134). Today, the indigenous population of Legal Amazonia (Brazil) is estimated at 175,000 people, 10 to 15 percent of which probably live in the cities.

The total population of Brazilian Amazon (Legal Amazonia) is about 24 million people, of which more than 3 million live in the two biggest Amazonian cities—Manaus and Belém—whose way of life does not differ significantly from that of most Western global cities. In rural settings, non-Amerindian[1] populations living off agroforestry resources such as açaí (*Euterpe oleracea*), hearts of palm, Brazil nuts, and other market-valued products, replicate pre-Columbian subsistence practices. In the Belterra Plateau and other areas, intensive, mechanized agriculture for exportation (especially soybeans), using pesticides and heavy machinery, coexists with extensive slash-and-burn farming (manioc, beans) and orchard cultivation (cacao [*Theobroma cacao*] and cupuaçu [*Theobroma grandiflorum*]), supplying local markets as well as being used for domestic consumption. The cultivation and processing of manioc (*Manihot esculenta*) into its many byproducts (mainly *farinha*, but also *tucupi* and tapioca) by *caboclos* and small, colonist farmers is believed to mirror pre-Columbian indigenous techniques, although the substitution of the stone axe for metal tools might have resulted in significant changes regarding the location of plots and the intensity of land use (Denevan 1992).

Cattle ranching, introduced in the region by the Jesuits in the eighteenth century, is believed to be one of the chief causes of present-day deforestation, although highway construction and colonization by small farmers compete in the statistics (Fearnside 2001a, 2007). Livestock farming today ranges from small herds owned by colonists, who have diversified their agribusiness to cope with crop losses and difficulties in reaching regional markets due to transportation problems, to large cattle industries owned by large companies that have launched Brazil as one of the largest beef exportation countries in the world.

Some small-scale, family-based economies developed in the region through the interregional migration of colonists brought in by settlement programs since the 1970s. The National Institute for Colonization and Agrarian Reform (INCRA) has distributed lots to immigrant families, which are quite different from native ones in terms of land-use strategies and cultural values. In some cases, they sell their lands to wealthier colonists or ranchers, who then use them for pasture or soybean cultivation (Fearnside 2001b, 2008). In the state of Acre, deforestation caused by these small agriculturalists surpasses forest razing brought about by logging and cattle ranching in many regions. This is because their plots are small, so the proportion of land they need to clear for cultivation is proportionally larger when compared with larger plots of land. As deforestation advances along the borders of the Amazon region (state of Acre, north of the state of Mato Grosso, south of the state of Pará, and west of the state of Maranhão), the "arc of deforestation" increases; satellite images show that, in huge degraded areas, indigenous territories (such as the Xingu Indigenous Park in the state of Mato Grosso and the Turiassu Indigenous Reservation in the state of Maranhão) appear as forest islands, oases that are testimonies of the evident link between indigenous practices and the survival of forest coverage (Balée, personal communication, 2010).

In the 1970s, a large number of immigrants were drawn to the Amazon region, along with the construction of highways that cut the forest from east to west and from south to north, resulting in higher rates of deforestation (Fearnside 2007). This process, however, began much earlier, with the rubber boom in the late nineteenth century (Brondizio 2006), continuing throughout the twentieth century. Waves of migrants formed a large, low-paid workforce that has continued to increase as mining industries were established in the Carajás Mine province, the Trombetas River, the city of Juruti (in the lower Amazon), and the state of Amapá, among other areas. Historical indigenous and African slave descendants' territories were encroached upon by the mining economy. Several small towns grew around the mining industries, creating a belt of poverty and social conflict.

Indigenous populations struggle to maintain their historical territories—many of them already guaranteed by law as *Terra Indígena*; (TI or Indigenous Territories)—in the face of invasions by *madeireiros* (lumbermen) and miners. Conservation areas created in the last decades may or may not have maintained traditional communities within its borders; having been pushed to the outer limits, some populations occasionally venture into those areas to collect the resources they used to have easy access to in the past. Extractive reserves created since the late 1980s could be an alternative for the sustainable use of forest products without causing deforestation. The government retains ownership of a land plot that is communally run by rubber tappers and Brazil nut or crab gatherers, among others. Fearnside (1989:388) sees this type of land use as promising, since the "proposal [by Brazil's National Council of Rubber Tappers and the Rural Worker's Union] originated at the grass-roots level," and is based on the maintenance of local techniques proven to be efficient for the removal of nontimber forest products with good market value.

Today, urban and rural Indian populations aggregate into new ethnic groups, claiming their indigenous identity and communal land rights. Given their traditional way of life, *quilombola* populations, sometimes considered riverine peoples, also have their own claim to territories and recognition; their rights are supported by the 1988 Constitution.

Extended family ties and community-based social relations survive in urban contexts generally marked by individualism and competition. Contemporary land use in Amazonia is the result of the manifold historicities of multiple peoples, cultures, and places, regional and national governmental policies, and their interconnectedness to the global village (see Brondizio 2006). Not only do Amazonian specificities have to be evaluated according to various analytical scales, considering how particular, regional situations may or may not contribute to the understanding of the big picture (Moran et al. 2008), but a long history of landscape manipulation also has to be taken into account for an accurate assessment of the present situation.

Deforestation has been a great concern for conservationists due to the alarming rates of forest coverage losses and its relation to global markets and interests. Forest burning (for cattle ranching and farming) has been one of the main worries of environmentalists, as it leads to an increase in deforestation and carbon dioxide emissions. In the light of historical ecology, however, fire is seen as a tool, a strategy humans have used throughout their history to "remake their environment, to transform it into forms more habitable and usable" (Pyne 1998:70). Forest burning is an old practice used to maintain savannas as open environments for farming and landscape engineering, to clear plots for cultivation, to cut down forests for village building, and to aid in hunting, pottery-making, and so on.

Unfortunately, distinguishing between anthropogenic and natural fires (caused by lightning) that occurred in prehistoric times is difficult (Pyne 1998:71). While excavating geoglyphs in Acre, we eventually found a charcoal-rich stratum some 10 to 20 centimeters below the surface, which we interpret as recent burning (in the last 20 years) for pasture. Along the Trombetas River, however, I have noticed a clear charcoal stratum below pre-Columbian cultural strata in excavations conducted by Vera Guapindaia, which seems to indicate that vast natural fires occurred more than 2,000 years ago. To my knowledge, this stratum has not been dated yet.

Savanna-burning in Llanos de Mojos has been described since colonial times, and landscape archaeological research shows evidence that burning was probably used well in pre-Columbian times as well. Erickson (2006:250) reports that present-day ranchers, farmers, and hunters burn the savannas because "fire removes dead grasses, encourages new grasses for live stock and game animals, keeps the forest at bay, and 'cleans' the landscape." On Marajó Island, hunters start fires to confine and easily catch small turtles known as *muçuãs* (*Kinosternon scorpioides*), desirable either for their own consumption or for sale.

Scholars are not sure what farming techniques were used by pre-Columbian Amerindians, since the long-fallow shifting cultivation, a mark of traditional societies in the region, might be a post-European development thanks to the availability of metal tools (Denevan 1992). However, evidence provided by the numerous sites containing a dark, highly fertile soil type (ADE) suggests that mulching and in-field burning was used to improve soil conditions (Denevan 2006). The addition of carbon to soils by Amazonians is still a poorly understood process, although they certainly produced charcoal by controlled burning, as this resulted in an incomplete combustion, leaving bits of charcoal intact, which helps maintain soil properties (nutrient levels, organic matter, high pH), and prevents leaching (Denevan 2006).

The Kyoto Protocol considers soils an important carbon storage site, and researchers investigating ADE soils believe they are a potential sink for carbon dioxide, or a carbon sequestration alternative (Sombroek et al. 2003). Landscapes have behaved differently according to the extension and type of fire produced, with different consequences for the global carbon cycle. Clearly, forest and savanna burning did not negatively affect the tropical lowlands before the European conquest, so the use of these techniques should not only be allowed to persist among traditional populations, but should also be investigated so as to learn where we can draw a line between ecological sustainable fires (those with the potential of increasing biomass and carbon sequestration) and those harmful to the biota. Pyne (1998:87) reserves the term "fire" for traditional open burning, calling the industrial use of fossil fuel "combustion." According to him, although industrial combustion substitutes traditional burning techniques, it cannot replace ancient fire ecology. Reducing or removing fire from landscapes will not restore a pristine nature, but "will probably create an environment that never before existed" (Pyne 1998:98).

Amazonia is important to the future of humanity. The conservancy of the region's biodiversity and the permanence of its vegetation coverage have been the cause of worldwide concern. Few have considered, however, the importance of the tropics in the sustainability of its own population, and how biological and cultural diversity are entangled. Unfortunately, social and cultural diversity have not been considered relevant to the conservation of the world's largest remaining tropical forest. Among other reasons, this is because the agency of Amazonian peoples is still eclipsed by a poor and biased view of the Amazonian environment as a pristine wilderness, and their inhabitants as "noble savages" (Balée 2008; Erickson and Balée 2006; Heckenberger 2005). A recent example is illustrative. In a TV commercial aired on Dia do Índio (Indian's Day) in Brazil (April 22, 2010), the vendor introduces a woman characterized as an indigenous female (using body painting and a feather headdress), and says: "To attest to this product, I've brought a legitimate representative of nature." In the Western imagination, indigenous peoples are part of nature, since supposedly they have always lived in symbiosis with the environment. Conversely, indigenous practices, documented through archaeological, historical, and ethnographic research, and forged through many centuries of active engagement with the natural world in coupled human-environment relations, are not considered valuable experiences for the adequate management of Amazonian resources.

Indigenous ways of life are part of the history of Amazonian landscapes. Ignoring this might jeopardize any efforts to preserve the tropical forest for future generations.

### NOTE

1. I use the term "non-Amerindian" to characterize populations that do not recognize themselves as indigenous, although they might be of indigenous descent. Alternatively, I use the term *caboclo*, which is not an ethnic category, but is largely used to name rural populations both with a negative and a positive connotation; in the latter case, when it is used to emphasize a way of life that represents indigenous heritage (Rodrigues 2006).

# References

Acevedo-Marín, Rosa, and Edna Maria Ramos de Castro
1998 *Negros do Trombetas: Guardiães de Matas e Rios*. Belém: Cejup.
Ackermann, Fritz L.
1963 O Lago Arari da Ilha de Marajó e Seus Problemas. *Revista Brasileira de Geografia Ano* XXV(2):273–276.
Adovasio, J. M., J. Donahue, and R. Stuckenrath
1990 The Meadowcroft Rockshelter Radiocarbon Chronology 1975–1990. *American Antiquity* 55:348–354.
Aihkenvald, Alexandra
1999 The Arawak Language Family. In *The Amazonian Languages*, R. W. Dixon and A. Aihkenvald, eds. Pp. 65–106. Cambridge: Cambridge University Press.
Almeida, Fernando Ozório de, and Lorena Gomes Garcia
2008 Aspectos do Espaço Tupinambá no Leste Amazônico. *Revista de Arqueologia* 21(2):97–119.
Alves, José de A., and José Seixas Lourenço
1981 Métodos Geofísicos Aplicados à Arqueologia no Estado do Pará. *Boletim do Museu Paraense Emílio Goeldi*. Série Geologia 26:1–52.
Ames, Kenneth M.
1995 Chiefly Power and Household Production on the Northwest Coast. In *Foundations of Social Inequality*, T. D. Price and G. M. Feinman, eds. Pp. 155–85. New York: Plenum Press.
Armentia, Padre Nicolas
1887 *Navegacion del Madre de Dios: Viaje del Padre Nicolas Armentia*. La Paz: Imprenta de La Paz.
Arnold, Dean E., and Kenneth A. Prettol
1988 Aboriginal Earthworks Near the Mouth of Beni, Bolivia. *Journal of Field Archaeology* 15(4):457–465.
Arnold, Jeanne E.
1996a *Emergent Complexity: The Evolution of Intermediate Societies*. Archaeological Series No. 9, Jeanne E. Arnold, ed. Ann Arbor: International Monographs in Prehistory.
1996b Organizational Transformations: Power and Labor among Complex Hunter-Gatherers and Other Intermediate Societies. In *Emergent Complexity: The Evolution of Intermediate Societies*. Archaeological Series No. 9, Jeanne E. Arnold, ed. Pp. 59–73. Ann Arbor: International Monographs in Prehistory.
Arroyo-Kalin, Manuel
2010 The Amazonian Formative: Crop Domestication and Anthropogenic Soils. *Diversity* 2(4):473–504.
Atwood, Frederick D.
1997 *Rocks and Minerals: A Portrait of the Natural World*. New York: Todtri.

Bailey, R. C., and T. N. Headland
1991 The Tropical Rain Forest: Is It a Productive Environment for Human Foragers? *Human Ecology* 19(2):261–285.

Balée, William
1989 The Culture of Amazonian Forests. In *Resource Management in Amazonia: Indigenous and Folk Strategies*. Advances in Economic Botany Volume 7, D. A. Posey and W. Balée, eds. Pp. 1–21. New York: New York Botanical Garden.
1998 *Advances in Historical Ecology* [editor]. New York: Columbia University Press.
2006 The Research Program of Historical Ecology. *Annual Review of Anthropology* 35:75–98.
2008 Sobre a Indigeneidade das Paisagens. *Revista de Arqueologia* 21(2):9–24.
2010 Contingent Diversity on Anthropic Landscapes. *Diversity* 2(2):163–181.

Balée, William, and Clark Erickson
2006 Time, Complexity, and Historical Ecology. In *Time and Complexity in Historical Ecology*, W. Balée and C. Erickson, eds. Pp. 1–17. New York: Columbia University Press.

Balser, Carlos
1974 *El Jade de Costa Rica: Un Albun Arqueologico*. San José, Costa Rica: Lehmann.

Bandeira, Arkley Marques
2008 Ocupações Humanas Pré-históricas no Litoral Maranhense: um Estudo Arqueológico Sobre o Sambaqui do Bacanga na Ilha de São Luís, Maranhão, Programa de Pós-Graduação em Arqueologia, USP.

Barata, Frederico
1950 *A Arte Oleira dos Tapajós: Considerações Sobre a Cerâmica e Dois Tipos de Vasos Característicos*. Instituto de Antropologia e Etnologia do Pará 2.
1951 *A Arte Oleira dos Tapajós II—Os Cachimbos de Santarém*. Revista do Museu Paulista NS V.
1953 *A Arte Oleira dos Tapajó III*. Alguns Elementos Novos para a Tipologia de Santarém, Volume 6. Belém: Instituto de Antropologia e Etnologia do Pará.

Barbosa Rodrigues, João de
1875 *Exploração e Estudo do Vale do Rio Amazonas*. Rio de Janeiro: Tipografia Nacional.
1899 *O Muyrakytã e os Ídolos Simbólicos*. Rio de Janeiro: Imprensa Nacional.

Barham, B. L., and O. T. Coones
1996 *Prosperity's Promise: The Amazon Rubber Boom and Distorted Development*. Boulder, CO: Westview Press.

Barse, William P.
1993 Review of Moundbuilders of the Amazon: Geophysical Archaeology on Marajó Island, Brazil, by Anna Curtenius Roosevelt. *American Antiquity* 58(2):373–374.

Barth, Fredrik
1969 *Ethnic Groups and Boundaries: The Social Organization of Culture Difference (Results of a Symposium Held at the University of Bergen, 23–26 February 1967)*. Bergen and London: Universitetsforlaget; Allen & Unwin.

Bates, Henry Walter
1892 *The Naturalist on the River Amazon*. London: Murray.

Beaudry, Mary, Lauren Cook, and Stephen Mrozowski
1991 Artifacts and Active Voices: Material Culture as Social Discourse. In *The Archaeology of Inequality*, R. H. McGuirre and R. Paynter, eds. Pp. 150–91. Cambridge, MA: Basil Blackwell.

Beltrão, Jane Felipe
  2008 Natureza de Nossa Vida! Barragem, Não Aceito! O Xingu é Sagrado. In *Histórias do Xingu: Fronteiras, Espaços e Territorialidades (Séculos XVII-XXI)*, C. M. d'Souza and A. Cardozo, eds. Pp. 205–214. Belém: Edufpa.
Bettendorf, Pe. João Felipe
  1990 *Crônica dos Padres da Companhia de Jesus no Estado do Maranhão*. Belém: Fundação Cultural do Pará Tancredo Neves, Secretaria de Estado da Cultura.
Bevan, Bruce W., and Anna C. Roosevelt
  2003 Geophysical Exploration of Guajará, a Prehistoric Earth Mound in Brazil. *Geoarchaeology: An International Journal* 18(3):287–331.
Binford, Lewis R.
  1971 Mortuary Practices: Their Study and Their Potential. *Memoirs of the Society for American Archaeology* 25:6–29.
Blanton, Richard, S. A. Kowalewski, and G. M. Feinman
  1981 *Ancient Mesoamerica: A Comparison of Change in Three Regions*. Cambridge: Cambridge University Press.
Boomert, Arie
  1987 Gifts of the Amazons: "Greenstone" Pendants and Beads as Items of Ceremonial Exchange in Amazonia. *Antropologica* 67:33–54.
Brochado, José Proenza
  1980 The Social Ecology of Marajoara Culture. M.A., Anthropology, University of Illinois.
  1984 An Ecological Model of the Spread of Pottery and Agriculture into Eastern South America. Ph.D. dissertation, University of Illinois at Urbana-Champaign.
Brondizio, Eduardo S.
  2006 Landscapes of the Past, Footprints of the Future: Historical Ecology and the Study of Contemporary Land-use Change in the Amazon. In *Time and Complexity in Historical Ecology*, W. Balée and C. L. Erickson, eds. Pp. 365–405. New York: Columbia University Press.
Brumfiel, Elizabeth M.
  1991 Weaving and Cooking: Women's Production in Aztec Mexico. In *Engendering Archaeology: Women and Prehistory*, J. Gero and M. W. Conkey, eds. Pp. 224–251. Cambridge: Blackwell.
Cabral, Mariana Petry, and João de Moura Saldanha
  2007 *Projeto de Investigação Arqueológica na Bacia do Rio Calçoene e Seu Entorno—Amapá*. 2° Relatório Semestral. Instituto de Pesquisas Científicas e Tecnológicas do Estado do Amapá.
  2008 Paisagens Megalíticas na Costa Norte do Amapá. *Revista de Arqueologia* 21(1):9–26.
Calado, Manuel
  2009 Resenha de Arqueologia da Amazônia Ocidental: Os Geoglifos do Acre. *Amazônica* 1(1):295–298.
Calandra, Horacio Adolfo, and Susana Alicia Salceda
  2004 Amazonia Boliviana: Arqueologia de los Llanos de Mojos. *Acta Amazônica* 34(2):155–163.
Caldarelli, Solange Bezerra, Fernanda de Araújo-Costa, and Dirse Clara Kern
  2005 Assentamentos a Céu Aberto de Caçadores-coletores Datados da Transição Pleistoceno Final/Holoceno Incial no Sudeste do Para. *Revista de Arqueologia* 18:95–108.

Carneiro, Robert L.
1960 Slash-and-Burn Agriculture: A Closer Look at Its Implications for Settlement Patterns. In *Men and Cultures*, A. F. C. Wallace, ed. Pp. 229–234. Philadelphia: University of Pennsylvania Press.
1961 Slash-and-Burn Cultivation among the Kuikuro and Its Implications for Cultural Development in the Amazon Basin. In *The Evolution of Horticultural Systems in Native South America: Causes and Consequences, a Symposium*, J. Wilbert, ed. Pp. 47–67. Caracas: Sociedade de Ciencias Naturales La Salla.
1970 A Theory of the Origin of State. *Science* 169:733–738.
1981 The Chiefdom: Precursor of the State. In *The Transition to Statehood in the New World*, G. Jones and R. Kautz, eds. Pp. 37–79. Cambridge: Cambridge University Press.
1983 The Cultivation of Manioc among the Kuikuro of the Upper Xingu. In *Adaptive Responses of Native Amazonians*, R. B. Hames and W. T. Vickers, eds. Pp. 65–111. New York: Academic Press.
1991 Chiefdom-level Warfare as Exemplified in Fiji and the Cauca Valley. In *The Anthropology of War*, J. Haas, ed. Pp. 190–211. Cambridge: Cambridge University Press.
1998 What Happened at the Flashpoint? Conjectures on Chiefdom Formation at the Very Moment of Conception. In *Chiefdoms and Chieftaincy in the Americas*, E. M. Redmond, ed. Pp. 18–42. Gainesville: University Press of Florida.
2007 A Base Ecológica dos Cacicados Amazônicos. *Revista de Arqueologia* 20:117–154.

Carvajal, Friar Gaspar de
1934 Discovery of the Orellana River. In *The Discovery of the Amazon According to the Account of Friar Gaspar de Carvajal and Other Documents*, J. T. Medina, ed. Pp. 167–235. New York: American Geographical Society.
2008 Relacion que Escribió Fr. Gaspar de Carvajal. In *La Aventura del Amazonas*, R. D. Maderuelo, ed. Pp. 31–88. Crónicas de América. Madrid: Dastín Export S.L.

Carvajal, Fray Gaspar de, P. de Almesto, and Alfonso de Rojas
2002 Relacion de Fray Gaspar de Carvajal. In *La Aventura del Amazonas*, R. D. Maderuelo, ed. Pp. 31–88. Crónicas de América. Spain: Dastín Export S. L.

Castello Branco, José Moreira Brandão
1960 *Descobrimento das Terras da Região Acreana*. Rio de Janeiro: Departamento de Imprensa Nacional.

Chambouleyron, Rafael
2008 O Sertão dos Taconhapé: Cravo, Índios e Guerras no Xingu Seiscentista. In *Histórias do Xingu: Fronteiras, Espaços e Territorialidades (Séculos XVII–XXI)*, C. M. d'Souza and A. Cardozo, eds. Pp. 51–71. Belém: Edufpa.

Chandless, William
1866 Ascent of the River Purus. *Journal of the Royal Geographical Society of London* 36:86–118.
1867 An Exploration of the River Aquiry, an Affluent of the Purus. *Proceedings of the Royal Geographical Society of London* 11(3):100–102.

Chernela, Janet M.
1982 Indigenous Forest and Fish Management in the Vaupés Basin of Brazil. *Cultural Survival Quarterly* 6(2):17–18.
1988 Righting History in the Northwest Amazon: Myth, Structure, and History in an Arapaço Narrative. In *Rethinking History and Myth: Indigenous South*

*American Perspectives on the Past*, J. D. Hill, ed. Pp. 35–49. Urbana and Chicago: University of Illinois Press.

1997 Pesca e Hierarquização Tribal no Alto Uaupés. In *Suma Etnológica Brasileira. Edição Atualizada do Handbook of South American Indians*, third edition, D. Ribeiro, ed. Pp. 279–295. Belém: Edufpa.

Church, George Earl
1877 The River Purus, in Its Commercial and Geographial Relations to the Valley of the Madeira. *Geographical Magazine* 4:95–99.

Clarke, Michael J.
2001 Akha Feasting: An Ethnoarchaeological Perspective. In *Feasts: Archaeological and Ethnographic Perspectives on Food, Politics, and Power*, M. Dietler and B. Hayden, eds. Pp. 144–167. Washington and London: Smithsonian Institution Press.

Clement, Charles R., M. De Cristo-Araújo, G. C. d'Eeckenbrugge, et al.
2010 Origin and Domestication of Native Amazonian Crops. *Diversity* 2(1):72–106.

Coirolo, Alicia Durán
2007 *Ilha de Terra: Uma Longa História*. Belém: Museu Paraense Emílio Goeldi [CD-ROM].

Corrêa, Conceição, Napoleão Figueiredo, and Mário F. Simões
1964 Projeto Marajó. Relatório de Excursão. Unpublished report, Belém: Museu Paraense Emílio Goeldi.

Costa, Angione
1941 Una Pieza de Culto Seual en la Arqueologia Amazonica. Congreso Internacional de Americanistas, Session 27, Lima, Peru: 1:297–304.

Costa, Jucilene Amorim da
2009 Geoquímica de Terra Preta Amazônica e suas Modificações Antrópicas no Vale do Rio Amazonas (Juruti, Pará). Ph.D. Dissertation. Department of Geosciences, Universidade Federal do Pará.

Costa, Marcondes L., Anna Cristina Resque Silva, and Rômulo S. Angélica
2002 Muyraquytã ou Muraquitã, um Talismã Arqueológico em Jade Procedente da Amazônia: uma Revisão Histórica e Considerações Antropogeológicas. *Acta Amazonica* 32(3):467–490.

Costa, Maria H. F., and Hamilton B. Malhano
1986 Habitação Indígena Brasileira. In *Suma Etnológica Brasileira, Volume 2—Tecnologia Indígena*, B. Ribeiro, ed. Pp. 27–94. Petrópolis: Vozes, Finep.

Costin, Cathy Lynne
1996 Exploring the Relationship between Gender and Craft in Complex Societies: Methodological and Theoretical Issues of Gender Attribution. In *Gender and Archaeology*, R. Wright, ed. Pp. 111–140. Philadelphia: University of Pennsylvania Press.

Crown, Patricia
2000 Gendered Tasks, Power, and Prestige in the Prehispanic Southwest. In *Women and Men in the Prehispanic Southwest: Labor, Power, and Prestige*, Patricia Crown, ed. Pp. 3–41. Santa Fe: School of American Research Press.

Crumley, Carole L.
1979 Three Locational Models: An Epistemological Assessment for Anthropology and Archaeology. In *Advances in Archaeological Method and Theory 2*. Pp. 141–175. New York: Academic Press.

1987 A Dialectical Critique of Hierarchy. In *Power Relations and State Formation*, T. C. Patterson and C. W. Gailey, eds. Pp. 155–169. Washington, D.C.: American Anthropological Association.
1994 Historical Ecology. A Multidimensional Ecological Orientation. In *Historical Ecology: Cultural Knowledge and Changing Landscapes*, C. L. Crumley, ed. Pp. 1–41. Santa Fe: School of American Research Press.
1995 Heterarchy and the Analysis of Complex Societies. In *Heterarchy and the Analysis of Complex Societies*, R. M. Ehrenreich, C. L. Crumley, and J. E. Levy, eds. Archaeological Papers of the American Anthropological Association No. 6.

Crumley, Carole, and William Marquardt
1990 Landscape: A Unifying Concept in Regional Analysis. In *Interpreting Space: GIS and Archaeology*, K. M. Allen, S. W. Gree, and E. B. Zubrow, eds. Pp. 73–79. London: Taylor and Francis.

Cruz, Friar Laureano de La
1900 *Nuevo Descobrimiento del Rio de Marañon Llamado de las Amazonas Hecho por Religión de San Francisco, Año de 1651*. Madrid: Biblioteca de la Irradiación.

Curet, L. Antonio, and Jose R. Oliver
1998 Mortuary Practices, Social Development, and Ideology in Precolumbian Puerto Rico. *Latin American Antiquity* 9(3):217–239.

D'Altroy, Terrence, and Timothy Earle
1985 Staple Finance, Wealth Finance, and Storage in the Inca Political Economy. *Current Anthropology* 26(2):187–206.

Daniel, Padre João
1840 *Tesouro Descoberto no Rio Amazonas*. Revista do Instituto Histórico e Geográfico Brasileiro 2.

Darvill, Timothy, and Julian Thomas
2001 *Neolithic Enclosures in Atlantic Northwest Europe, Volume 6*. Oxford: Oxbow Books.

Davis, Christopher
2009 Archaeoastronomy at Monte Alegre, Pará: A Research Problem and Research Strategy. *Amazonica* 1(2):537–547.

de Acuña, C.
1859 A New Discovery of the Great River of the Amazons. In *Expeditions into the Valley of the Amazons*, C. R. Markham, ed. Pp. 44–142. London: Hakluyt Society.

DeBoer, Warren R.
1975 The Archaeological Evidence for Manioc Cultivation: A Cautionary Note. *American Antiquity* 40:419–433.
1998 Figuring Figurines: The Case of the Chachi, Ecuador. In *Recent Advances in the Archaeology of Northern Andes*, A. Oyuela-Caycedo and J. S. Raymond, eds. Los Angeles: University of California.

Denevan, William M.
1963 Additional Comments on the Earthworks of Mojos in Northeastern Bolivia. *American Antiquity* 28(4):540–545.
1966 *The Aboriginal Cultural Geography of the Llanos de Mojos*. Berkeley: University of California Publications.
1976 The Aboriginal Population of Amazonia. In *The Native Populations of the Americas before 1492*, William Denevan, ed. Madison: University of Wisconsin Press.
1980 *La Geografia Cultural Aborigen de los Llanos de Mojos*. Las Paz: Juventud.

1992 Stone vs Metal Axes: The Ambiguity of Shifting Cultivation in Prehistoric Amazonia. *Journal of the Steward Anthropological Society* 20:153–165.
1996 A Bluff Model of Riverine Settlement in Prehistoric Amazonia. *Annals of the American Association of Geographers* 86(4):654–681.
2001 *Cultivated Landscapes of Native Amazonia and the Andes: Triumph over the Soil.* Oxford and New York: Oxford University Press.
2003 The Native Population of Amazonia in 1492 Reconsidered. *Revista de Indias* LXIII(227):175–188.
2006 Pre-European Forest Cultivation in Amazonia. In *Time and Complexity in Historical Ecology: Studies in the Neotropical Lowlands*, W. Balée and C. L. Erickson, eds. Pp. 153–163. New York: Columbia University Press.
2009 Introduction to Indian Adaptations in Flooded Regions of South America. *Journal of Latin American Geography* 8(2):209–224.

Derby, Orville
1879 The Artificial Mounds of the Island of Marajo. *American Naturalist* 13:224–229.

Descola, Philippe
1994 [1986] *In the Society of Nature: A Native Ecology in Amazonia.* Cambridge: Cambridge University Press.

Dias, Adriana Schmidt
1995 Um Projeto para a Arqueologia Brasileira: Breve Histórico da Implementação do PRONAPA. *Revista do CEPA* 19(22):25–39.

Dias, Ondemar
1977 *Relatório do Primeiro Año de Pesquisas no Estado do Acre.* Rio de Janeiro: IAB—MPEG/PRONAPABA.
2006 Arqueologia da Amazônia Ocidental—Descrição Sumária das Características da Tradição Quinari (Alto Curso do Rio Purus). In *Estudos Contemporâneos de Arqueologia*, O. Dias, E. Carvalho, and M. Zimmermann, eds. Pp. 169–205. Palmas: Unitins: IAB.

Dias, Ondemar, and Eliana Carvalho
1988 *As Estruturas de Terra da Arqueologia do Acre.* Rio de Janeiro: Série ArqueoIAB, Publicações Avulsas 1.

Dillehay, Tom D., and David J. Meltzer
1991 *Peopling of the New World: Problems, Processes and Prospects.* Boca Raton, FL: CRC Press.

Donatti, Patricia B.
2003 A Ocupação Pré-colonial da Área do Lago Grande, Iranduba, AM. Master's thesis, Department of Archaeology, University of São Paulo, Museum of Archaeology and Ethnology.

Drennan, Robert D.
1995 Mortuary Practices in the Alto Magdalena: The Social Context of the "San Augustín Culture." In *Tombs for the Living: Andean Mortuary Practices*, T. D. Dillehay, ed. Washington, D.C.: Dumbarton Oaks.

Duarte-Filho, Aiezer
2010 Do Rio Nhamundá ao Amazonas e Tapajós: uma Rota Transveral Pré-colonial na Região do Baixo Amazonas. In *Monografia de Especialização em Arqueologia*. Belém: Universidade Federal do Pará.

Dunnell, Robert C.
1992 The Notion Site. In *Space, Time, and Archaeological Landscapes*, J. Rossignol and L. Wandsnider, eds. Pp. 21–42. New York: Plenum Press.

Earle, Timothy
 1977 A Reappraisal of Redistribution: Complex Hawaiian Chiefdoms. In *Exchange Systems in Prehistory*, T. Earle and J. Ericson, eds. Pp. 213–29. New York: Academic Press.
 1990 Style and Iconography as Legitimation in Complex Chiefdoms. In *The Uses of Style in Archaeology*, M. Conkey and C. Hastorf, eds. Pp. 73–81. Cambridge: Cambridge University Press.
 1997 *How Chiefs Come to Power: The Political Economy in Prehistory*. Stanford: Stanford University Press.
Easby, Elizabeth K.
 1968 *Pre-Columbian Jade from Costa Rica*. New York: André Emmerich Inc.
Eder, Francisco J.
 1985 *Breve Descripcion de las Reducciones de Mojos ca. 1772*, J. M. Barnadas, transl. Cochabamba, Bolivia: Historia Boliviana.
Edmundson, G.
 1920 The Voyage of Pedro Teixeira on the Amazon from Pará to Quito and Back, 1637–1639. *Transactions of the Royal Historical Society* 3:52–71.
Eidt, R. C.
 1985 Theoretical and Practical Considerations in the Analysis of Anthrosols. In *Archaeological Geology*. New Haven and London: Yale University Press.
Erickson, Clark L.
 1980 Sistemas Agrícolas Prehispanicos en los Llanos de Mojos. *América Indígena* 40(4):731–755.
 1995 Archaeological Methods for the Study of Ancient Landscapes of the Llanos de Mojos in the Bolivian Amazon. In *Archaeology in the Lowland American Tropics*, P. Stahl, ed. Pp. 66–95. Cambridge: Cambridge University Press.
 2000 An Artificial Landscape-scale Fishery in the Bolivian Amazon. *Nature* 408:190–193.
 2001 Pre-Columbian Fish Farming in the Amazon. *Expedition* 43(3):7–8.
 2003 Historical Ecology and Future Explorations. In *Amazonian Dark Earths: Explorations in Space and Time*, B. Glaser and W. Woods, eds. Pp. 455–500. Berlin: Springer-Verlag.
 2004 Historical Ecology and Future Explorations. In *Amazonian Dark Earths: explorations in space and time*, B. Glaser and W. Woods, eds. Pp. 455–500. Berlin: Springer-Verlag.
 2006 The Domesticated Landscapes of the Bolivian Amazon. In *Time and Complexity in Historical Ecology*, W. Balée and C. Erickson, eds. Pp. 235–278. New York: Columbia.
 2008 Amazonia: The Historical Ecology of a Domesticated Landscape. In *Handbook of South American Archaeology*, H. Silvermann and W. Isbell, eds. Pp. 157–183. New York: Springer.
 2010 The Transformation of Environment into Landscape: The Historical Ecology of Monumental Earthwork Construction in the Bolivian Amazon. *Diversity* 2(1):618–652.
Erickson, Clark, Patricia Alvarez, and Sergio Calla
 2008 Zanjas Circundantes: Obras de Tierra Monumentales de Baures en la Amazonia Boliviana. Proyecto Agro-Arqueológico del Beni. Informe del Trabajo de Campo de la Temporada 2007. Unpublished.
Erickson, Clark, and William Balée
 2006 The Historical Ecology of a Complex Landscape in Bolivia. In *Time and Complexity in Historical Ecology: Studies in the Neotropical Lowlands*, W. Balée and C. L. Erickson, eds. Pp. 187–234. New York: Columbia University Press.

Eriksen, Love
  2011 *Nature and Culture in Prehistoric Amazonia: Using G.I.S. to Reconstruct Ancient Ethnogenetic Processes from Archaeology, Linguistics, Geography, and Ethnohistory*. Lund Studies in Human Ecology 12. Lund: Media-Tryck, Lund University.
Estrada, Emilio
  1961 *Nuevos Elementos en la Cultura Valdivia: Sus Posibles Contactos Transpacíficos*. Guayaquil: Publicacion del Sub-Comite Ecuatoriano de Antropologia, Instituto Panamericano de Geografia Y Historia.
Evans, Clifford, and Betty Jane Meggers
  1965 *Guia para Prospecção Arqueológica no Brasil*. Belém: CNPq, INPA, MPEG.
Farabee, William Curtis
  1921 Explorations at the Mouth of the Amazon. *Museum Journal* 12:142–161.
Fawcett, Percy Harrison
  2001 *Exploration Fawcett*. London: Phoenix Press.
Fearnside, Philip M.
  1989 Extractive Reserves in Brazilian Amazonia: An Opportunity to Maintain Tropical Rain Forest under Sustainable Use. *BioScience* 39(6):387–393.
  2001a La Deforestación en Amazonia. In *Amazonia: Orientaciones para el Desarollo Sostenible*, M. Pyhälä and S. F. Paitan, eds. Pp. 61–71. Lima: Embajada de Finlandia.
  2001b Soybean Cultivation as a Threat to the Environment in Brazil. *Environmental Conservation* 28(1):23–38.
  2007 Brazil's Cuiabá-Santarém (BR-163) Highway: The Environmental Cost of Paving a Soybean Corridor through the Amazon. *Environmental Management* 39:601–614.
  2008 Comment: Will Urbanization Cause Deforested Areas to be Abandoned in Brazilian Amazonia? *Environmental Conservation* 35(3):197–199.
Ferrarini, Sebastião Antonio
  2009 *Rio Purus: História, Cultura, Ecologia*. São Paulo: FTD.
Ferreira, Alexandre Rodrigues
  1974 *Viagem Filosófica pelas Capitanias do Grão Pará, Rio Negro, Mato Grosso e Cuiabá*. Conselho Federal de Cultura.
Ferreira Penna, Domingos Soares
  1876 Breve Notícia Sobre os Sambaquis do Pará. *Archivos do Museu Nacional do Rio de Janeiro* 1:85–99.
  1877 Apontamentos Sobre os Cerâmios do Pará. *Archivos do Museu Nacional do Rio de Janeiro* 2:47–67.
  1885 Índios de Marajó. *Archivos do Museu Nacional do Rio de Janeiro* 6:108–115.
Figueiredo, Napoleão
  1963 Projeto Marajó: Relatório de Excursão. Unpublished fieldwork report, Belém: Museu Paraense Emílio Goeldi.
Figueiredo, Napoleão, and Mário F. Simões
  1962 Projeto Marajó: Relatório de Excursão. Unpublished fieldwork report, Belém: Museu Paraense Emílio Goeldi.
Flannery, Kent
  1976 Linear Stream Pattern and Riverside Settlement Rules. In *The Early Mesoamerican Village*, K. Flannery, ed. Pp. 173–180. New York: Academic Press.
Ford, James
  1949 Cultural Dating of Prehistoric Sites in Viru Valley, Peru. In *Surface Survey of the Viru Valley, Peru*. American Museum of Natural History, Anthropological Papers 43:31–78.

Freidel, David A.
1993 The Jade Ahau: Toward a Theory of Commodity Value in Maya civilization. In *Precolumbian Jade: New Geological and Cultural Interpretations*, F. W. Lange, ed. Pp. 149–165. Salt Lake City: University of Utah Press.

Gallo, Giovanni
1996 *O Homem que Implodiu*. Belém: Secult.

Garson, A.
1980 Prehistory, Settlement and Food Production in the Savanna Region of La Calzada de Paez, Venezuela. Ph.D. dissertation, Department of Anthropology, Yale University.

Geertz, Clifford
1983 *Local Knowledge: Further Essays in Interpretive Anthropology*. New York: Basic Books.

Gennep, Arnold van
1977 *Rites of Passage*. London: Routledge.

Gilman, Antonio
1976 Bronze Age Dynamics in Southeast Spain. *Dialectical Anthropology* 1:307–319.
1991 Trajectories Towards Social Complexity in the Later Prehistory of the Mediterranean. In *Chiefdoms: Power, Economy and Ideology*, T. Earle, ed. Pp. 146–168. Cambridge: Cambridge University Press.

Ginsburg, Faye D., and Rayna Rapp [eds.]
1995 *Conceiving the New World Order: The Global Politics of Reproduction*. Berkeley: University of California Press.

Glaser, B., J. Lehmann, and W. Zech
2002 Ameliorating Physical and Chemical Properties of Highly Weathered Soils in the Tropics with Charcoal: A Review. *Biology and Fertility of Soils* 35:219–230.

Gomes, Denise M. C.
2001 Santarém: Symbolism and Power in the Tropical Forest. In *Unknown Amazon: Culture in Nature in Ancient Brazil*. C. McEwan, C. Barreto, and E. Neves, eds. Pp. 134–155. London: British Museum Press.

Gomes, Denise Maria Cavalcante
1997 Bibliografia Sobre a Cultura Santarém: História e Perspectivas. *Revista do Museu de Arqueologia e Etnologia* 7:155–166.
2005 Análise Dos Padrões de Organização Comunitária no Baixo Tapajós: o Desenvolvimento do Formativo na Área de Santarém, PA. Ph.D. dissertation, University of São Paulo.
2006 Padrões de Organização Comunitária no Baixo Tapajós: o Formativo na Área de Santarém, Brasil. In *Pueblos y Paisajes Antiguos de la Selva Amazónica*, G. Morcote, S. Mora, and C. Franky, eds. Pp. 237–254. Bogotá and Washington, D.C.: Taraxacum.
2007 The Diversity of Social Forms in Pre-colonial Amazonia. *Revista de Arqueologia Americana* 25:189–226.
2010 Os Contextos e os Significados da Arte Cerâmica dos Tapajó. In *Arqueologia Amazônica 1*, E. Pereira and V. Guapindaia, eds. Belém: MPEG.

Gott, Richard
1993 *Land Without Evil, Utopian Journeys across the South American Watershed*. London, New York: Verso.

Green, Lesley, David Green, and Eduardo Góes Neves
2003 Indigenous Knowledge and Archeological Science. *Journal of Social Archaeology* 3(3):365–397.

Guapindaia, Vera
  1993 Fontes Históricas e Arqueológicas Sobre os Tapajó: a Coleção Frederico Barata do Museu Paraense Emílio Goeldi. Master's thesis, Federal University of Pernambuco.
  2001 Encountering the Ancestors: The Maracá Urns. In *Unknown Amazon: Culture in Nature in Ancient Brazil*, C. McEwan, C. Barreto, and E. Neves, eds. London: The British Museum Press.
  2009 Além da Margem do Rio—a Ocupação Konduri e Pocó na Região de Porto Trombetas, PA. Ph.D. dissertation, Department of Archaeology, University of São Paulo.
Guerrero, Juan Vicent
  1998 The Archaeological Context of Jade in Costa Rica. In *Jade in Ancient Costa Rica*, J. Jones, ed. Pp. 39–58. New York: Metropolitan Museum of Art.
Guzmán, Décio de Alencar
  2008 O Inferno Abreviado: Evangelização e Expansão Portuguesa no Xingu (Século XVII). In *Histórias do Xingu: Fronteiras, Espaços e Territorialidades (Séculos XVII–XXI)*, C. M. d'Souza and A. Cardozo, eds. Pp. 35–49. Belém: Edufpa.
Hartt, Charles Frederick
  1871 The Ancient Indian Pottery of Marajo, Brazil. *American Naturalist* 5(5):259–271.
  1876 Notas Sobre Algumas Tangas de Barro Cozido dos Antigos Indígenas da Ilha de Marajó. *Archivos do Museu Nacional do Rio de Janeiro 1*.
  1885 Contribuições para a Ethnologia do Valle do Amazonas. *Archivos do Museu Nacional do Rio de Janeiro VI*.
Hastorf, Christine A.
  1991 Gender, Space, and Food in Prehistory. In *Engendering Archaeology: Women and Prehistory*, J. Gero and M. W. Conkey, eds. Pp. 132–159. Cambridge, MA: Blackwell.
Heckenberger, Michael J.
  1996 War and Peace in the Shadow of Empire: Sociopolitical Change in the Upper Xingu of Southeastern Amazonia A.D. 1400–2000. Ph.D. dissertation, University of Pittsburgh.
  1998 Manioc Agriculture and Sedentism in Amazonia: The Upper Xingu Example. *Antiquity* 72:633–648.
  1999 O Enigma das Grandes Cidades: Corpo e Estado na Amazonia. In *A Outra Margem do Ocidente (Brasil 500 Años: Experiência e Destino)*, A. Novaes, ed. Pp. 125–152. São Paulo: Cia das Letras.
  2001a Estrutura, História e Transformação: a Cultura Xinguana na Long Durée, 1000-2000 d.C. In *Povos do alto Xingu: História e Cultura*, M. J. Heckenberger and B. Franchetto, eds. Pp. 21–62. Rio de Janeiro: Universidade Federal do Rio de Janeiro.
  2001b Rethinking the Arawakan Diaspora: Hierarchy, Regionality, and the Amazonian "Formative." In *Comparative Arawak Histories*, J. N. Hill and F. Santos-Granero, eds. Urbana-Champaign: University of Illinois Press.
  2005 *The Ecology of Power: Culture, Place, and Personhood in the Southern Amazon, A.D. 1000-2000*. New York and London: Routledge.
  2010 Biocultural Diversity in the Southern Amazon. *Diversity* 2(1):1–16.
Heckenberger, Michael J., Afukaka Kuikuro, Urissapá Tabata Kuikuro, et al.
  2003 Amazonia 1492: Pristine Forest or Cultural Parkland? *Science* 301:1710–1713.

Heidegger, Martin
    1973 Art and Space. *Man & World* 6:3–8.
Helms, Mary W.
    1979 *Ancient Panama: Chiefs in Search of Power.* Austin and London: University of Texas Press.
    1998 *Access to Origins: Affines, Ancestors and Aristocrats.* Austin: University of Texas Press.
Hemming, J.
    1978 *Red Gold: The Conquest of the Brazilian Indians.* London: Macmillian London Ltd.
Heriarte, Mauricio de
    1874 [1662] *Descripção do Estado do Maranhão, Pará, Corupá e Rio das Amazonas.* Vienna: Imprensa do Filho de Carlos Gerold.
Hertz, R.
    1960 *Death and the Right Hand.* London: Cohen & West.
Hilbert, Klaus, and Peter Paul Hilbert
    1980 Resultados Preliminares da Pesquisa Arqueológica nos Rios Nhamundá e Trombetas. *Boletim do Museu Paraense Emílio Goeldi.* Série Antropologia 75.
Hilbert, Peter Paul
    1952 Contribuição à Arqueologia da Ilha de Marajó: Os Tesos Marajoaras do Alto Camutins e a Atual Situação da Ilha do Pacoval, no Arari. *Instituto de Antropologia e Etnologia do Pará* 5:5–32.
Hilbert, Peter Paul, and Klaus Hilbert
    1979 Archäologische Untersuchungen am Rio Nhamundá, Unterer Amazonas. *Beiträge zur Allgemeinen und Vergleichenden Archäologie Band* 1:439–450.
Hodder, Ian
    1982 Sequences of Structural Change in the Dutch Neolithic. In *Symbolic and Structural Archaeology,* Ian Hodder, ed. Pp. 162–178. Cambridge: Cambridge University Press.
    2005 An Archaeology of the Four-field Approach in Anthropology in the United States. In *Unwrapping the Sacred Bundle: Reflections on the Disciplining of Anthropology,* D. A. Segal and S. J. Yanagisako, eds. Pp. 126–140. Durham: Duke University Press.
Huntington, R., and P. Metcalf
    1991 *Celebrations of Death: The Anthropology of Mortuary Ritual.* Cambridge and New York: Cambridge University Press.
Iriarte, José, Óscar Marozzi, and Christopher Gillam
    2010 Monumentos Funerários y Festejos Rituales: Complejos de Recintos y Túmulos Taquara/Itararé en Eldoado, Misiones (Argentina). *Arqueologia Iberoamericana* 6:25–38.
Isaac, Barry L.
    1975 Resource Scarcity, Competition, and Cooperation in Cultural Evolution. In *A Reader in Cultural Change, Volume I: Theories,* I. A. Brady and B. L. Isaac, eds. Pp. 125–143. New York: Schenkman.
Jackson, Jean
    1994 Becoming Indians: The Politics of Tukanoan Ethnicity. In *Amazonian Indians from Prehistory to the Present: Anthropological Perspectives,* A. C. Roosevelt, ed. Pp. 383–406. Tucson: The University of Arizona Press.
Johnson, Allen W., and Timothy K. Earle
    2000 *The Evolution of Human Societies: From Foraging Group to Agrarian State.* Stanford, CA: Stanford University Press.

Johnson, Gregory A.
   1972 A Test of the Utility of Central Place Theory in Archaeology. In *Man, Settlement, and Urbanism*, P. J. Ucko, R. Tringham, and G. W. Dimbleby, eds. Pp. 769–785. London: Duckworth.
   1977 Aspects of Regional Analysis in Archaeology. *Annual Review of Anthropology* 6:479–508.
Junker, Laura Lee, Karen Mudar, and Marla Schwaller
   1994 Social Stratification, Household Wealth, and Competitive Feasting in 15th/16th-century Philippine Chiefdoms. *Research in Economic Anthropology* 15:307–358.
Jurandir, Dalcídio
   1994 *Três Casas e um Rio*. Belém: CEJUP.
Kern, Dirse Clara
   1988 Caracterização Pedológica de Solos com Terra Preta Arqueológica na Região de Oriximiná. Master's thesis, Department of Agronomy, Federal University of Rio Grande do Sul, Brazil.
   1996 Geoquímica e Pedogeoquímica em Sítios Arqueológicos com Terra Preta na Floresta Nacional de Caxiuanã (Portel/PA) Ph.D. dissertation, Post-graduate Course in Geology and Geochemistry, Federal University of Para.
Kern, Dirse C., and Nestor Kampf
   1989 Antigos Assentamentos Indígenas na Formação de Solos com Terra Preta Arqueológica na Região de Oriximiná, Pará. *Revista Brasileira de Ciência do Solo* 13:219–225.
Kern, Dirse C., M. L. Costa, J. L. Frazão, and M. Jardim
   1999 A Influência das Palmeiras como Fonte de Elementos Químicos em Sítios Arqueológicos com Terra Preta. Paper presented at the VI Simpósio de Geologia da Amazônia, Belém, Brazil, June 15–17.
Kern, Dirse C., M. L. P. Ruivo, and F. J. L. Frazao
   2009 Terra Preta Nova: The Dream of Wim Sombroek. In *Amazonian Dark Earth: Wim Sombroek's Vision*, William I. Woods, Wenceslau G. Teixeira, Johannes Lehmann, Christoph Steiner, Antoinette WinklerPrins, and L. Rebellato, eds. Pp. 339–349. Berlin: Springer.
Kipnis, Renato, Solange Caldarelli, and Wesley Charles Oliveira
   2005 Contribuição para a Cronologia da Colonização Amazônica e Suas Implicações Teóricas. *Revista de Arqueologia* 18:81–93.
Labre, Colonel A. R. Pereira
   1888 Viagem Exploradora do Rio Madre de Dios ao Acre. *Revista de Geografia do Rio de Janeiro* IV(2):102–116.
   1889 Colonel Labre's Explorations in the Region between the Beni and Madre de Dios Rivers and the Purus. *Proceedings of the Royal Geographical Society and Monthly Record of Geography* 11(8):496–502.
Lathrap, Donald W.
   1970a La Floresta Tropical y el Contexto Cultural de Chavín. In *100 Años de Arqueología en el Perú*, R. Ravines, ed. Fuentes e Investigaciones para la Historia del Perú, 3. Lima: Instituto de Estudios Peruanos: Edición de Petróleos del Perú.
   1970b *The Upper Amazon*. New York: Praeger.
   1973 The Antiquity and Importance of Long-distance Trade Relationship in the Moist Tropics of Pre-Columbian South America. *World Archaeology* 5:170–186.
   1985 Jaws: The Control over Power in the Early Nuclear American Ceremonial Center. In *Early Ceremonial Architecture in the Andes*, C. B. Donnan, ed. Pp. 241–268. Washington, D.C.: Dumbarton Oaks.

Lehmann, J., C. V. Campos, J. L. V. Macedo, and L. German
   2003 Sequential P Fractionation and Sources of P in Amazonian Dark Earths. In *Explorations in Amazonian Dark Earths,* B. Glaser and W. I. Woods, eds. Pp. 113–23. Berlin: Springer.
Lévi-Strauss, Claude
   1993 Rousseau, Fundador das Ciências do Homem. In *Antropologia Estrutural 2.* P. 41. Rio de Janeiro: Tempo Brasileiro.
Lima, Anderson Marcio Amaral
   2010 Compreendendo Estruturas de Piso. Unpublished report, Belém: Universidade Federal do Pará.
Lima, André da Silva
   2006 A Guerra Pelas Almas: Alianças, Recrutamento e Escravidão Indígena—do Maranhão ao Cabo do Norte, 1615–1647. Pós Graduação em História Social da Amazônia, UFPA.
Lima, Luiz Fernando Erig
   2003 Levantamento Arqueológico das Áreas de Interflúvio na Área de Confluência dos Rios Negro e Solimões, AM. Master's thesis, Graduate Program in Archaeology, University of São Paulo.
   2010 Pedras Verdes, Piedras Hijadas ou Spleen Stones: o Comércio de Pedras na Amazônia Indígena sob o Olhar dos Europeus. *Amazônica* 2(2):298–313.
Linné, Sigwald
   1928 Les Recherches Archéologiques de Nimuendajú au Brésil. *Journal de la Société des Americanistes de Paris (Nouvelle Série)* XX:71–91.
Lopes, Paulo R. C.
   1999 A Colonização Portuguesa da Ilha de Marajó: Espaço e Contexto Arqueológico-histórico na Missão Religiosa de Joanes. Master's thesis, History Graduate Program, Pontifícia Universidade Católica do Rio Grande do Sul.
López-Mazz, José M.
   2001 Las Estructuras Tumulares (Cerritos) del Litoral Atlântico Uruguayo. *Latin American Antiquity* 12(3):231–255.
Lowie, Robert H.
   1948 The Tropical Forests: An Introduction. In *Handbook of South American Indians,* J. Steward, ed. Pp 1–2. Bureau of American Ethnology, Bulletin 143. Washington, D.C.: Smithsonian Institution.
Machado, Juliana Salles
   2005 Montículos Artificiais na Amazônia Central: um Estudo de Caso do Sítio Hatahara. Master's thesis, University of São Paulo.
Magalhães, Marcos
   2005 *A Physis da Origem: o Sentido da História na Amazônia.* Belém: MPEG.
Magalis, Joanne E.
   1975 A Seriation of Some Marajoara Painted Anthropomorphic Urns. Ph.D. dissertation, University of Illinois.
Marajó, Barão de
   1895 *As Regiões Amazônicas—Estudos Corográficos dos Estados do Grão-Pará e Amazonas.* Lisboa: Libâneo da Silva.
Marquardt, William H.
   1992 *Culture and Environment in the Domain of the Calusa.* Gainesville: University of Florida Press.
Martins, Cristiane Maria Pires, A. Marcio Amaral Lima, Denise P. Schaan et al.
   2010 Padrões de Sepultamento na Periferia do Domínio Tapajó. *Amazonica* 2(1):137–139.

Maurtua, Victor M.
　1906 *Juicio de los Limites entre el Peru y Bolivia*. 12 volumes. Barcelona: Henrich Y Co.
McAnany, Patricia A.
　1995 *Living with the Ancestors: Kinship and Kingship in Ancient Maya Society*. Austin: University of Texas Press.
McCann, John M., William I. Woods, and D. W. Meyer
　2000 Organic Matter and Anthrosols in Amazonia: Interpreting the Amerindian Legacy. Paper presented at the British Society of Soil Science Proceedings, London.
McEwan, Colin
　2001 Seats of Power: Axiality and Access to Invisible Worlds. In *Unknown Amazon: Nature and Culture in Ancient Brazil*, C. McEwan, C. Barreto, and E. Neves, eds. Pp. 176–195. London: British Museum Press.
Meggers, Betty J.
　1945 The Beal-Steere Collection of Pottery from Marajó Island. *Papers of the Michigan Academy of Science, Arts and Letters* XXXI(3).
　1954 Environmental Limitation on the Development of Culture. *American Anthropologist* 56(5):801–824.
　1971 *Amazonia: Man and Culture in a Counterfeit Paradise*. Chicago: Aldine Atherton.
　1985 Advances in Brazilian Archaeology, 1935–1985. *American Antiquity* 50(2):364–373.
　1992 Review of Moundbuilders of the Amazon: Geophyiscal Archaeology on Marajó Island, Brazil. *Journal of Field Archaeology* 19:399–404.
　1995 Judging the Future by the Past: The Impact of Environmental Stability on Prehistoric Amazonian Populations. In *Indigenous Peoples and the Future of Amazonia: An Ecological Anthropology of an Endangered World*, L. E. Sponsel, ed. Pp. 15–43. Tucson and London: The University of Arizona Press.
　1998 Jomon-Valdivia Similarities: Convergence or Contact? In *Across before Columbus? Evidence for Transoceanic Contact with the Americas Prior to 1492*, D. Y. Gilmore and L. S. McElroy, eds. Pp. 11–22. Edgecomb, ME: The New England Antiquities Research Association.
　2001 The Mystery of the Marajoara: An Ecological Solution. *Amazoniana* XVI(3/4):421–40.
Meggers, Betty J., and Jacques Danon
　1988 Identification and Implications of a Hiatus in the Archeological Sequence on Marajo Island, Brazil. *Journal of Washington Academy of Sciences* 78(3):245–253.
Meggers, Betty J., and Clifford Evans
　1957 Archeological Investigations at the Mouth of the Amazon. *Bureau of American Ethnology Bulletin 167*. Washington, D.C.: Smithsonian Institution.
　1961 An Experimental Formulation of Horizon Styles in the Tropical Forest Area of South America. In *Essays in Pre-Columbian Art and Archaeology*, S. K. Lothrop, ed. Pp. 372–388. Cambridge: Harvard University Press.
　1970 *Como Interpretar a Linguagem da Ceramica: Manual para Arqueologos*. Washington D.C.: Smithsonian Institution.
Meggers, Betty J., Clifford Evans, and Emilio Estrada
　1965 *Early Formative Period of Coastal Ecuador: the Valdivia and Machalilla Phases*. Washington, D.C.: Smithsonian Institution.
Meltzer, David J.
　1988 Late Pleistocene Human Adaptations in Eastern North America. *Journal of World Prehistory* 2(1):1–52.

Métraux, Alfred
1942 *The Native Tribes of Eastern Bolivia and Western Matto Grosso.* Washington, D.C.: Smithsonian Institution.

Metraux, Alfred
1947 Mourning Rites and Burial Forms of the South American Indians. *America Indígena* 7:7–44.

Morais, Claide de Paula
2006 Arqueologia na Amazônia Central Vista a Aprtir de uma Perspectivada Região do Lago do Limão. Master's thesis, Graduate Program in Archaeology. University of São Paulo.

Morales-Chocano, Daniel
2000 Las Poblaciones Prehistóricas Amazónicas. *Investigaciones Sociales* IV(6):71–92.

Morán, Emilio F.
1995 Disaggregating Amazonia: A Strategy for Understanding Biological and Cultural Diversity. In *Indigenous People and the Future of Amazonia,* L. E. Sponsel, ed. Pp. 71–95. Tucson and London: The University of Arizona Press.

Moran, Emilio F., Eduardo S. Brondízio, and Mateus Batistella
2008 Trajetórias de Desmatamento e uso da Terra na Amazônia Brasileira: uma Análise Multiescalar. In *Amazônia: Natureza e Sociedade em Transformação,* M. Batistella, E. F. Moran, and D. S. Alves, eds. Pp. 55–70. São Paulo: Edusp.

Morcote-Rios, Gaspar
2006 Plantas y Gentes Antiguas en un Igapó Estacional del Interfluvio Solimões-Içá (Amazonas-Putumayo). In *Pueblos y Paisajes Antiguos en la Selva Amazónica,* G. Morcote-Rios, S. Mora-Camargo, and C. Franky-Calvo, eds. Pp. 256–259. Bogotá: Taraxacum.

Mordini, Antonio
1929 Couvre Sexe Precolombiani in Argilla dell'Isola di Marajo. *Archivio per L'Anthropologia e la Etnolografia* 59.
1936 Contributo allo Studio dell'Archeologia dell'Isola di Marajo. *Il Nazionale, Revista di Studi Americani Ano* XIV(479).

Munn, Nancy D.
1962 Walbiri Graphic Signs: An Analysis. *American Anthropologist* 64:972–984.
1973 The Spatial Presentation of Cosmic Order in Walbiri Iconography. In *Primitive Art and Society,* A. Forge, ed. Pp. 193–220. London: Oxford University.

Myers, Thomas P.
1973 Toward a Reconstruction of Prehistoric Community Patterns in the Amazon Basin. In *Variation in Anthropology: Essays in Honor of John C. McGregor,* D. Lathrap and J. Douglas, eds. Pp. 233–252. Urbana: Illinois Archaeological Survey.
1981 Hacia la Reconstruccion de los Patrones Comunales de Asentamiento durante la Prehistoria de la Cuenca Amazonica. *Amazonia Peruana* IV(7):31–63.

Myers, Thomas, William Denevan, Antoinette WinklerPrins, and Antonio Porro
2003 Historical Perspectives on Amazonian Dark Earths. In *Amazonian Dark Earths: Origin, Properties, Management,* J. Lehmann, D. C. Kern, B. Glaser, and W. I. Woods, eds. Pp. 15–28. Dordrecht: Kluwer Academic.

Netto, Ladislau de Souza Mello
1885 Investigações Sobre a Arqueologia Brasileira. *Archivos do Museu Nacional do Rio de Janeiro* 6:257–554.

Neves, Eduardo G.
1998 Paths in Dark Waters: Archaeology as Indigenous History in the Upper Rio Negro Basin, Northwest Amazon. Ph.D. dissertation, Department of Anthropology, Indiana University.
2008 Ecology, Ceramic Chronology and Distribution, Long-term History, and Political Change in the Amazonian Floodplain. In *Handbook of South American Archaeology*, H. Silverman and W. H. Isbell, eds. Pp. 359–379. New York: Springer.
Neves, Marcos Vinicius
2002 História Nativa do Acre: Povos do Acre. In *História Indígena da Amazônia Ocidental*. Pp. 10–15. Rio Branco: Cimi/FEM.
Nimuendajú, Curt
1949 Os Tapajó. *Boletim do Museu Paraense Emílio Goeldi* 10:93–106.
2002 *Mapa Etno-Histórico de Curt Nimuendajú*. Rio de Janeiro: IBGE.
2004 *In Pursuit of a Past Amazon: Archaeological Researches in the Brazilian Guyana and in the Amazon Region*. Gotenborg: Elanders Infologistik.
Noelli, Francisco S.
1996 As Hipóteses Sobre o Centro de Origem e Rotas de Expansão dos Tupi. *Revista de Antropologia* 39(2):7–118.
Nordenskiöld, Erland
1930 *L'Archaeologie du Basin de L'Amazone*. Paris: G. van Oest.
2009 Indian Adaptations in Flooded Regions of South America. *Journal of Latin American Geography* 8(2):209–224.
Noronha, José Monteiro de
2006 *Roteiro da Viagem da Cidade do Pará até as Últimas Colônias do Sertão da Província (1768)*. São Paulo: Edusp.
Nunes Filho, Edinaldo Pinheiro
2005 *Pesquisa Arqueológica no Amapá*. Macapá: B-A-Bá.
O'Dwyer, Eliane Cantarino
2006 Quilombos e as Fronteiras da Antropologia. In *Os Caminhos do Patrimônio no Brasil*, M. F. L. Filho and M. Bezerra, eds. Pp. 105–123. Goiânia: Alternativa.
Oberg, Karlevo
1973 [1955] Types of Social Structure among the Lowland Tribes of South and Central America. In *Peoples and Cultures of Native South America*, Volume 3, D. R. Gross, ed. Pp. 189–212. New York: Doubleday, The Natural History Press.
Okada, H.
1998 Maritime Adaptations in Northern Japan. *Artic Anthropology* 35(1):335–339.
Organização dos Estados Americanos (OEA)
1974 *Marajó. Um Estudo para seu Desenvolvimento*. Washington, D.C.: Organização dos Estados Americanos.
Oswald, Alastair, Carolyn Dyer, and Martyn Barber
2001 *The Creation of Monuments: Neolithic Causewayed Enclosures in the British Isles*. Swindon: English Heritage.
Otten, Charlotte
1971 *Anthropology and Art: Readings in Cross-Cultural Aesthetics*. Austin: University of Texas Press.
Oyewùmí, Oyèrónkâe
1997 *The Invention of Women: Making an African Sense of Western Gender Discourses*. Minneapolis: University of Minnesota Press.

Pacheco, Agenor Sarraf
2009 História e Literatura no Regime das Águas: Práticas Culturais Afro-indígenas na Amazônia Marajoara. *Amazonica* 1(2):406–441.
Palmatary, Helen C.
1939 *Tapajó Pottery.* Volume 8. Göteborg: Göteborgs Etnografiska Museum.
1950 The Pottery of Marajo Island, Brazil. *Transactions of the American Philosophical Society* 39(3).
1960 The Archaeology of the Lower Tapajós Valley, Brazil. *Transactions of the American Philosophical Society* 50(3).
Parsons, James J., and William A. Bowen
1966 Ancient Ridged Fields of the San Jorge River Floodplain, Colombia. *The Geographical Review* 56(3):317–343.
Pärssinen, Martti, Alceu Ranzi, Sanna Saunaluoma, and Ari Siiriäinen
2003 Geometrically Patterned Ancient Earthworks in the Rio Branco Region of Acre, Brazil: New Evidence of Ancient Chiefdom Formations in Amazonian Interfluvial Terra Firme Environment. *Renvall Institute Publications, University of Helsinki* 14:97–133.
Pärssinen, Martti, Denise P. Schaan, and Alceu Ranzi
2009 Pre-Columbian Geometric Earthworks in the Upper Purus: A Complex Society in Western Amazonia. *Antiquity* 83(322):1084–1095.
Pärssinen, Martti, and Denise Pahl Schaan
2005 *Natureza e Sociedade na História da Amazônia Ocidental. Projeto de Pesquisa.* Belém: MPEG, University of Helsinki.
Pärssinen, Martti, Ari Siiriäinen, and Antti Korpisaari
2003b Fortifications Related to the Inca Expansion. In *Western Amazonia—Amazônia Ocidental. Multisdisciplinary Studies on Ancient Expansionst Movements, Fortifications and Sedentary Life,* M. Pärssinen and A. Korpisaari, eds. Pp. 29–72. Helsinki: Renvall Institute Publications, University of Hensinki.
Pereira, Edithe
2003 *Arte Rupestre na Amazônia—Pará.* São Paulo: Unesp.
Perota, Celso
1992 Adaptação Agrícola no Baixo Xingu. In *Prehistoria Sudamericana. Nuevas Perspectivas,* Betty Meggers, ed. Pp. 211–218. Washington, D.C.: Taraxacum.
Plafker, George
1963 Observations on Archaeological Remains in Northeastern Bolivia. *American Antiquity* 28(3):372–378.
Plazas, Clemencia, A. M. Falchetti, J. Sáenz, and S. Archila
1993 *La Sociedad Hidráulica Zenú: Estudio Arqueológico de 2000 Años de Historia en las Llanuras del Caribe Colombiano.* Santafé de Bogotá: Editorial Banco de la República.
Politis, Gustavo
2001 Foragers of the Amazon: The Last Survivors or the First to Suceeed? In *Unknown Amazon: Culture in Nature in Ancient Brazil,* C. McEwan, C. Barreto, and E. Neves, eds. Pp. 26–49. London: The British Museum Press.
2007 *Nukak: Ethnoarchaeology of an Amazonian People.* Walnut Creek, CA: Left Coast Press.
Porras, P. I.
1987 *Investigaciones Archaeológicas a las Faldas de Sangay.* Quito: Centro de Investigaciones Arqueológicas.

Porro, Antonio
 1994 Social Organization and Political Power in the Amazon Floodplain: The Ethnohistorical Sources. In *Amazonian Indians from Prehistory to the Present: Anthropological Perspectives*, Anna C. Roosevelt, ed. Pp. 79–94. Tucson: University of Arizona Press.
 1996 *O Povo das Águas: Ensaios de Etno-história Amazônica*. Petrópolis, São Paulo: Vozes, Edusp.
 2010 Arte e Simbolismo Xamânico na Amazônia. *Boletim do Museu Paraense Emílio Goeldi. Ciências Humanas* 5(1):129–144.
Posey, Darrell A.
 1985 Indigenous Management of Tropical Forest Ecosystems: The Case of the Kayapó Indians of the Brazilian Amazon. *Agroforestry Systems* 3:139–158.
Pouguet, Martial
 2002 *Chronologie de la Période Céramique de l'Achéologie Amazonienne: Réflexions Théoriques et Méthodologiques*. Pontifícia Universidade Católica do Rio Grande do Sul.
Prance, G., and T. Lovejoy, eds.
 1985 *Key Environments: Amazonia*. New York: Pergamon Press.
Pyne, Stephen J.
 1998 Forged in Fire: History, Land, and Anthropogenic Fire. In *Advances in Historical Ecology*, W. Balée, ed. Pp. 64–103. New York: Columbia University Press.
Quinn, Ellen R.
 2004 Excavating "Tapajó" Ceramics at Santarém: Their Age and Archaeological Context. Ph.D. dissertation, Department of Anthropology, University of Illinois at Chicago.
Ranzi, Alceu
 2003 Geoglifos. Patrimônio Cultural do Acre. *Renvall Institute Publications, University of Helsinki* 14:135–172.
 2008 *Paleontologia da Amazônia. Mamíferos Fósseis do Juruá*. Rio Branco: M. M. Paim.
Ranzi, Alceu, and Rodrigo Aguiar
 2000 Registro de Geoglifos na Região Amazônica—Brasil. *Munda* 42:87–90.
 2004 *Geoglifos da Amazônia—Perspectiva Aérea*. Florianópolis: Faculdades Energia.
Ranzi, Alceu, Roberto Feres, and Foster Brown
 2007 Internet Software Programs Aid in Search for Amazonian Geoglyphs. *Eos* 88(21–22):226,229.
Raymond, J. Scott
 1995 From Potsherds to Pots: A First Step in Constructing Cultural Context from Tropical Forest Archaeology. In *Archaeology in the Lowland American Tropics: Current Analytical Methods and Applications*, P. W. Stahl, ed. Pp. 224–242. Cambridge: Cambridge University Press.
Rebellato, Lilian
 2007 Interpretando a Variabilidade Cerâmica e as Asinaturas Químicas e Físicas do Solo no Sítio Arqueológico Hatahara, AM. Master's thesis, Graduate Program in Archaeology, University of São Paulo.
Redman, Charles L.
 1999 *Human Impact on Ancient Environments*. Tucson: The University of Arizona Press.
Reich, Alfred, and Stegelmann, Felix
 1903 Bei den Indianern des Urubamba und dos Envira. *Globus* 83(9):133–137.

Reichel-Dolmatoff, Gerardo
1961 Anthropomorphic Figurines from Colômbia, Their Magic and Art. In *Essays in Precolumbian Art and Archaeology*, S. Lothrop, ed. Pp. 229–241. Cambridge: Harvard University Press.
1971 *Amazonian Cosmos*. Chicago: University of Chicago Press.
Renfrew, Colin
1976 Megaliths, Territories and Populations. In *Aculturation and Continuity in Atlantis Europe*, S. DeLaet, ed. Pp. 198–220. Bugres: de Tempel.
1982 Socio-economic Change in Ranked Societies. In *Ranking, Resource, and Exchange: Aspects of the Archaeology of Early European Society*, C. Renfrew and S. Shennan, eds. Pp. 1–8. Cambridge and New York: Cambridge University Press.
1986 Introduction: Peer Polity Interaction and Socio-political Change. In *Peer Polity Interaction and Sociopolitical Change*, C. Renfrew and J. F. Cherry, eds. Pp. 1–18. Cambridge: Cambridge University Press.
Ribeiro, Berta
1987 A Linguagem Simbólica da Cultural Material: Introdução. In *Suma Etnológica Brasileira, Volume 3*, B. Ribeiro, ed. Petrópolis: Vozes/FINEP.
Richardson III, James B.
1994 *People of the Andes*. Washington D.C.: Smithsonian Institution.
1998 Looking in the Right Places: Pre-5,000 B.P. Maritime Adaptations in Peru and the Changing Environment. *Revista de Arqueologia Americana* 15:33–56.
Rodrigues, Aryon Dall'Igna
2001 Biodiversidade e Diversidade Etnolingüística na Amazônia. In *Cultura e Biodiversidade entre o Rio e a Floresta*, M. d. S. Simões, ed. Pp. 269–278. Belém: Edufpa.
Rodrigues, Carmen Isabel
2006 Caboclos na Amazônia: a Identidade na Diferença. *Novos Cadernos NAEA* 9(1):119–130.
Rogez, Hervê
2000 *Açaí: Preparo, Composição e Melhoramento da Conservação*. Belém: Edufpa.
Roosevelt, Anna C.
1980 *Parmana: Prehistoric Maize and Manioc Subsistence Along the Amazon and Orinoco*. New York: Academic Press.
1988 Interpreting Certain Female Images in Prehistoric Art. In *The Role of Gender in Precolumbian Art and Architecture*, V. E. Miller, ed. Pp. 1–34. Chicago: University Press of America.
1991 *Moundbuilders of the Amazon: Geophysical Archaeology on Marajo Island, Brazil*. San Diego: Academic Press.
1995 Early Pottery in the Amazon: Twenty Years of Scholarly Obscurity. In *The Emergence of Pottery*, W. K. Barnett and J. W. Hoopes, eds. Pp. 115–131. Washington, D.C. and London: Smithsonian Institution.
1999 The Development of Prehistoric Complex Societies: Amazonia: A Tropical Forest. In *Complex Polities in the Ancient Tropical World*, E. A. Bacus and L. J. Lecero, eds. Pp. 13–33.
2000 The Lower Amazon: A Dynamic Human Habitat. In *Imperfect Balance: Landscape Transformations in the Precolumbian Americas*, D. L. Lentz, ed. Pp. 455–492. New York: Columbia University Press.
Roosevelt, Anna C., R. A. Housle, Maura I. da Silveira, et al.
1991 Eighth Millenium Pottery from a Prehistoric Shell Midden in the Brazilian Amazon. *Science* 254:1557–1696.

1996 Paleoindian Cave Dwellers in the Amazon: The Peopling of America. *Science* 272:372–384.
Rostain, Stéphen
2010 Pre-Columbian Earthworks in Coastal Amazonia. *Diversity* 2:331–352.
Rusby, Henry H.
1933 *Jungle Memories*. New York and London: Whittlesey House, McGraw-Hill.
Russell, Pamela
1998 The Palaeolithic Mother-Goddess: Fact or Fiction? In *Reader in Gender Archaeology*, K. Hays-Gilpin and D. Whitley, eds. Pp. 261–268. London: Routledge.
Rye, Owen S.
1981 *Pottery Technology: Principles and Reconstruction*. Manuals on Archaeology 4. Washington, D.C.: Taraxacum.
Sahlins, Marshall D.
1958 *Social Stratification in Polynesia*. Seattle: University of Washington Press.
1972 *Stone Age Economics*. Chicago: Aldine.
Salazar, Ernesto
2000 *Pasado Precolombino de Morona Santiago*. Macas, Ecuador: Casa de la Cultura Ecuatoriana Benjamin Carrión.
2008 Pre-Columbian Mound Complexes in the Upano River Valley, Lowland Ecuador. In *Handbook of South American Archaeology*, H. Silverman and W. Isbell, eds. Pp. 263–278. New York: Springer.
Sandweiss, Daniel H., Heather McInnins, Richard L. Burger, et al.
1998 Quebrada Jaguay: Early South American Maritime Adaptations. *Science* 281(5384):1830–1832.
Santos-Granero, Fernando
2004 Arawakan Sacred landscapes: Emplaced Myths, Place Rituals, and the Production of Locality in Western Amazonia. In *Kultur, Raum, Landschaft: Zur Bedeutung des Raumes in Zeiten der Globalität*, E. Halbmayer and E. Mader, eds. Pp. 93–122. Frankfurt: Brandes and Apsel Verlag.
Sauer, Carl Ortwin
1969 The Morphology of Landscape (1925). In *Land and Life: A Selection from the Writings of Carl Ortwin Sauer*, J. Leighly, ed. Pp. 315–350. Berkeley/Los Angeles: University of California Press.
Saunaluoma, Sanna
2010 Pre-Columbian Earthworks in the Riberalta Region of the Bolivian Amazon. *Amazonica* 2(1):86–115.
Scatamacchia, Maria Cristina M.
2000 *Mostra do Redescobrimento*. São Paulo: Fundação Bienal de São Paulo/ Associação Brasil 500 Años Artes Visuais.
Schaan, Denise P.
1997 *A Linguagem Iconográfica da Cerâmica Marajoara. Um Estudo da Arte Pré-histórica na Ilha de Marajó, Brasil (400-1300 AD)*. Porto Alegre: Edipucrs.
1999 Evidências para a Permanência da Cultura Marajoara à Época do Contato Europeu. *Revista de Arqueologia* 12/13:23–42.
2001a Estatuetas Marajoara: o Simbolismo de Identidades de Gênero em uma Sociedade Complexa Amazônica. *Boletim do Museu Paraense Emílio Goeldi. Série Antropologia* 17(2):23–63.
2001b Os Dados Inéditos do Projeto Marajó (1962–1965). *Revista do Museu de Arqueologia e Etnologia, São Paulo* 11:141–164.
2001c Into the Labyrinths of Marajoara Pottery: Status and Cultural Identity in an Amazonian Complex Society. In *The Unknown Amazon: Nature in Culture*

*in Ancient Brazil,* C. McEwan, C. Barreto, and E. G. Neves, eds. Pp. 108–133. London: British Museum Press.
2003 Investigating Gender in Prehistoric Amazonia. In *Women's Studies News,* P. 4. Pittsburgh: University of Pittsburgh.
2004 The Camutins Chiefdom: Rise and Development of Complex Societies on Marajó Island, Brazilian Amazon. Ph.D. dissertation, Department of Anthropology, University of Pittsburgh.
2006 Diagnóstico do Patrimônio Arqueológico na Área de Influência da Rodovia BR-163: Trecho Santarém-Rurópolis. Unpublished report, Belém: Universidade Federal do Pará.
2007 Múltiplas Vozes, Histórias e Memórias: por uma Gestão Compartilhada do Patrimônio Arqueológico da Amazônia. *Revista do Patrimônio: Arqueologia* (32):19–35.
2008 The Nonagricultural Chiefdoms of Marajó Island. In *Handbook of South American Archaeology,* H. Silverman and W. Isbell, eds. Pp. 339–357. New York: Springer.
2009a *Cultura Marajoara.* Rio de Janeiro: Senac Editoras, Federação de Comércio do Estado do Pará, Centro do Comércio do Pará.
2009b *Cultura Marajoara/Marajoara Culture* [Trilingual edition]. Rio de Janeiro: SENAC.
2010a Long-Term Human Induced Impacts on Marajó Island Landscapes, Amazon Estuary. *Diversity* 2(2):182–206.
2010b *Salvamento do Sítio PA-ST-42: Porto de Santarém. Relatório Final.* UFPA. Belém: JKNoronha.
Schaan, Denise P., M. Bueno, A. Ranzi, et al.
2010a Construindo Paisagens como Espaços Sociais: o Caso dos Geoglifos do Acre. *Revista de Arqueologia* 23(1):30–41.
Schaan, Denise P., Dirse Clara Kern, and Francisco Frazão
2009a An Assessment of the Cultural Practices Behind the Formation (or Not) of Anthropogenic Black Earth in Marajo Island Archaeological Sites. In *Amazonian Dark Earths: Wim Sombroek's Vision,* W. Woods, W. Teixeira, J. Lehmann, C. Steiner, A. WinklerPrins, and L. Rebellato, eds. Pp. 127–141. Berlin: Springer.
Schaan, Denise P., and Cristiane Pires Martins (eds.)
2010 *Muito Além dos Campos: Arqueologia e História na Amazônia Marajoara.* Belém: GKNoronha
Schaan, Denise P., Martti Pärssinen, Alceu Ranzi, et al.
2007 Geoglifos da Amazônia Ocidental: Evidência de Complexidade Social Entre Povos da Terra Firme. *Revista de Arqueologia* 20:67–82.
Schaan, Denise P., M. Pärssinen, S. Saunaluoma, et al. n.d.
New Radiometric Dates for Pre-Columbian (2000-700 B.P.) Earthworks in Western Amazonia, Brazil. *Journal of Field Archaeology* (submitted).
Schaan, Denise Pahl, and Wagner Fernando da Veiga e Silva
2004 O Povo das Águas e sua Expansão Territorial: uma Abordagem Regional de Sociedades Pré-coloniais na Ilha de Marajó. *Revista de Arqueologia* 17:13–32.
Schaan, Denise P., Maura Silveira, Fernando Marques, et al.
2009b Arqueologia da Costa Atlântica Amazônica: Síntese e Perspectivas. Unpublished manuscript.
Scheel-Ybert, Rita, Caroline F. Caromano, Leandro M. Cascon, et al.
2010 Estudos de Paleoetnobotânica, Paleoambiente e Paisagem na Amazônia Central e o Exemplo do Sudeste-sul do Brasil. In *Arqueologia Amazônica, Volume 2,* E. Pereira and V. Guapindaia, eds. Pp. 909–935. Belém: MPEG.

Schiffer, Michael B.
1987 *Formation Processes of the Archaeological Record*. Albuquerque, NM: University of New Mexico Press.
Schmidt, Morgan J., and Michael J. Heckenberger
2009 Amerindian Anthrosols: Amazonian Dark Earth Formation in the Upper Xingu. In *Amazonian Dark Earths: Wim Sombroek's Vision*, W. Woods, W. Teixeira, J. Lehmann, C. Steiner, A. WinklerPrins, and L. Rebellato, eds. Pp. 163–191. Berlin: Springer.
Schmidt, Morgan J., and A. G. Noack
2000 Black Carbon in Soils and Sediments: Analysis, Distribution, Implications, and Current Challenges. *Global Biochemical Cycles* 14:777–793.
Segal, Daniel Alan, and Sylvia Junko Yanagisako
2005 *Unwrapping the Sacred Bundle: Reflections on the Disciplining of Anthropology*. Durham: Duke University Press.
Service, Elman R.
1962 *Primitive Social Organization: An Evolutionary Perspective*. New York: Random House.
Shennan, Stephen
1982 Exchange and Ranking: The Role of Amber in the Earlier Bronze Age of Europe. In *Ranking, Resource, and Exchange: Aspects of the Archaeology of Early European Society*, Colin Renfrew, ed. Pp. 33–45. Cambridge: Cambridge University Press.
Shepard, Anna O.
1956 *Ceramics for the Archaeologist*. Washington D.C.: Carnegie Institution of Washington.
Silva, Fabíola Andréa
2002 Mito e Arqueologia: a Interpretação dos Asurini do Xingu Sobre os Vestígios Arqueológcos Encontrados no Parque Indígena Kuatinemu, Pará. *Horizontes Antropológicos* 8(18):175–187.
2009 Arqueologia e Etnoarqueologia na Aldeia Lalima e na Terra Indígena Kayabi: Reflexões Sobre Arqueologia Comuitária e Gestão do Patrimônio Arqueológico. *Revista do Museu de Arqueologia e Etnologia* 19:205–219.
2010 A Etnoarqueologia na Amazônia: Contribuições e Perspectivas. *Boletim do Museu Paraense Emílio Goeldi* 4(1):27–37.
Silva, Tallyta Suenny Araujo da In Press
Tecnologia Lítica na Amazônia: Análise das Lâminas de Machados Encontrados ao Longo da BR-163: Cuiabá-Santarém. Notícia de Pesquisa em Andamento. *Amazonica* 3(1).
Silveira, Maura Imazio
1994 Estudo Sobre Estratégia de Subsistência de Caçadores-coletores Pré-históricos do Sítio Gruta do Gavião, Carajás, PA. Master's thesis, University of São Paulo.
Silveira, Maura Imazio, Maria Christina L. Rodrigues, Elisangela Oliveira, et al.
2008 Seqüencia Cronológica de Ocupação na Área do Salobo (Pará). *Revista de Arqueologia* 21(1):61–84.
Silveira, Maura Imazio, and Denise P. Schaan
2005 Onde a Amazônia Encontra o Mar: Estudando os Sambaquis do Pará. *Revista de Arqueologia* 18:67–79.
Simões, Mário F.
1967 Resultados Preliminares de uma Prospecção Arqueológica na Região dos Rios Goiapi e Camará (Ilha de Marajó). In *Atas do Simpósio Sobre a Biota Amazônica*, Herman Lent, ed. Pp. 207–224. Belém: CNPq.

1969 The Castanheira Site: New Evidence on the Antiquity and History of the Ananatuba Phase (Marajó Island, Brazil). *American Antiquity* 34(4):402–410.
1981 Coletores-Pescadores Ceramistas do Litoral do Salgado (Pará). Nota Preliminar. *Boletim do Museu Paraense Emílio Goeldi. Série Antropologia* 78.
Simões, Mário F., and Napoleão Figueiredo
1965 Projeto Marajó. Final Report. Unpublished report, Belém: Museu Paraense Emílio Goeldi.
Smith, Nigel
1980 Anthrosols and Human Carrying Capacity in Amazonia. *Annals of the American Association of Geographers* 70(4):553–566.
2001 Land Use Dynamics in the Amazon Estuary and Implications for Natural Resource Management. Amazoniana XVI(3/4):517–537.
2002 *Amazon Sweet Sea: Land, Life, and Water at the River's Mouth*. Austin: University of Texas Press.
Sombroek, William G.
1966 *Amazon Soils: A Reconnaissance of the Soils of the Brazilian Amazon Region*. Wageningen: Centre for Agricultural Publication and Documentation.
Sombroek, Wim, M. L. Ruivo, P. M. Fearnside, B. Glaser, and J. Lehmann
2003 Amazonian Dark Earths as Carbon Stores and Sinks. In *Amazonian Dark Earths: Origins, Properties, Management*, J. Lehmann, D. Kern, B. Glaser, and W. Woods, eds. Pp. 125–139. Dordrecht: Kluwer Academic Publishers.
Sousa, Greyce Pereira de Oliveira e
2008 Projeções do Cotidiano: uma Análise dos Vestígios Materiais de um Passado Recente no Município de Juruti, Pará. Unpublished report, Belém: UFPA.
Spencer, Charles, and Elsa Redmond
1998 Prehispanic Causeways and Regional Politics in the Llanos de Barinas, Venezuela. *Latin American Antiquity* 9(2):95–110.
Spencer, Charles S., Elsa M. Redmond, and Milagro Rinaldi
1994 Drained Fields at la Tigra, Venezuelan Llanos: A Regional Perspective. *Latin American Antiquity* 5(2):119–43.
Stenborg, Per, Denise P. Schaan, and Anderson Marcio Amaral Lima n.d.
Precolumbian Land Use and Settlement Pattern in the Santarém Region, Lower Amazon. Latin American Antiquity (submitted).
Steward, Julian H.
1948 The Tropical Forest Tribes. In *Handbook of South American Indians, Volume 3*, J. Steward, ed. Bureau of American Ethnology Bulletin 143. Washington D.C.: Smithsonian Institution.
1949 The Native Population of South America. In *Handbook of South American Indians*, J. Steward, ed. Pp. 507–533. Bureau of American Ethnology Bulletin 143. Washington, D.C.: Smithsonian Institution.
Steward, Julian H., and Louis C. Faron
1959 *Native Peoples of South America*. London: McGraw-Hill Book Company.
Steward, Julian H., and Alfred Métraux
1948 Tribes of the Peruvian and Ecuadorian Montaña. In *Handbook of South American Indians, Volume 3: The Tropical Forest Tribes*, J. Steward, ed. Pp. 535–657. Bureau of American Ethnology Bulletin 143. Washington D.C.: Smithsonian Institution.
Stone-Miller, Rebecca
1995 *Art of the Andes: From Chavín to Inca*. London: Thames and Hudson.

Teles, Eliana, and Rosa Elizabeth Acevedo-Marin
  2009 Ecologia e Uso de Recursos dos Quilombolas dos Rios Arari e Gurupa, Ilha de Marajó, PA. Paper presented at XIV Ciso—Encontro de Ciências Sociais do Norte e Nordeste. Recife, September 8–11.
Tessmann, Günter
  1928 *Menschen ohne Gott.* Stuttgart: Sttecker und Schröeder.
  1930 *Die Indianer Nordost-Perus.* Hamburg: Friederichsen, de Gruyter & Co.
Tibón, Gutierre
  1983 *El Jade de Mexico: El Mundo Esotérico del "Chalchihuite."* Mexico City: Panorama Editorial.
Tocantins, Sylvia Helena
  2005 *A Lenda do Amor Eterno: Romance Marajoara.* Belém: Impr. Oficial do Pará.
Torres, Heloisa A.
  1940 *A Arte Indígena da Amazônia.* Publicações do SPHAN No. 6. Rio de Janeiro: Imprensa Nacional.
Tsai, S. M., B. O'Neill, F. S. Cannavan, D. Saito, N. P. S. Falcao, D. C. Kern, J. Grossman, and J. Thies
  2009 The Microbial World of Terra Preta. In *Amazonian Dark Earths: Wim Sombroek's Vision,* W. Woods, W. Teixeira, J. Lehmann, C. Steiner, A. WinklerPrins, and L. Rebellato, eds. Pp. 299–308. Berlin: Springer.
Turner, Terence
  1980 The Social Skin. In *Not Work Alone: A Cross-Cultural View of Activities Superfluous to Survival,* Jeremy Cherfas and and Roger Lewin, eds. Pp. 112–140. London: Temple Smith.
Turner, Victor Witter
  1967 *The Forest of Symbols: Aspects of Ndembu Ritual.* Ithaca, NY: Cornell University Press.
  1974 *Dramas, Fields, and Metaphors: Symbolic Action in Human Society.* Ithaca, NY: Cornell University Press.
Ucko, Peter J.
  1968 *Anthropomorphic Figurines of Predynastic Egypt and Neolithic Crete.* London: Andrew Szmidla.
Velthem, Lúcia Hussak van
  1994 Arte Indígena: Referentes Sociais e Cosmológicos. In *Índios no Brasil,* L. D. Grupioni, ed. Brasília: Min. da Educação e do Desporto.
  1998 *A Pele de Tuluperê: uma Etnografia dos Trançados Wayana.* Belém: Museu Paraense Emílio Goeldi.
Vianna, João
  1998 *A Fazenda Aparecida: Romance.* Belém: SECULT-PA.
Vidal, Lux
  1992 *Grafismo Indígena: Estudos de Antropologia Estética.* São Paulo: Studio Nobel/Fapesp/Edusp.
Vieira, Pe. Antônio
  1992 *Escritos Instrumentais Sobre os Índios.* São Paulo: EDUC/Loyola/Giordano.
Villas-Boas, Orlando, and Cláudio Villas-Boas
  1985 *Xingu: os Índios, Seus Mitos.* Porto Alegre: Kuarup.
Virtanen, Pirjo K.
  2010 Constancy in Continuity? Native Oral History, Iconography, and Earthworks on the Upper Purus River. Unpublished manuscript.

Viveiros de Castro, Eduardo
  1987 A Fabricação do Corpo na Sociedade Xinguana. In *Sociedades Indígenas e Indigenismo no Brasil*, O. Filho, ed. Rio de Janeiro: Marco Zero.
  1996a Images of Nature and Society in Amazonian Ethnology. *Annual Review of Anthropology* 25:179–200.
  1996b Os Pronomes Cosmológicos e o Perspectivismo Ameríndio: Mana. *Estudos de Antropologia Social* 2(2):115–144.
Walker, John H.
  2008 Pre-Columbian Ring Ditches Along the Yacuma and Rapulo Rivers, Beni, Bolivia: A Preliminary Review. *Journal of Field Archaeology* 33(4):413–427.
Wassén, S. Henry
  1934 The Frog-Motive among the South American Indians. *Anthropos* 29(3–4):319–337.
Whitehead, Neil
  1994 The Ancient Amerindian Polities of the Amazon, the Orinoco, and the Atlantic Coast: A Preliminary Analysis of Their Passage from Antiquity to Extinction. In *Amazonian Indians from Prehistory to the Present: Anthropological Perspectives*, Anna C. Roosevelt, ed. Pp. 33–53. Tucson: University of Arizona Press.
Widmer, Randolph J.
  1988 *The Evolution of the Calusa: A Nonagricultural Chiefdom on the Southwest Florida Coast*. Tuscaloosa: University of Alabama Press.
Wiessner, Pauline W.
  1983 Style and Social Information in Kalahari San Projectile Points. *American Antiquity* 48:253–276.
Wiessner, Polly
  1989 Style and Changing Relations between the Individual and Society. In *The Meanings of Things: Material Culture and Symbolic Expression*, Ian Hodder, ed. Pp. 56–63. London and Boston: Unwin Hyman.
Wobst, H. Martin
  2000 Agency in (Spite of) Material Culture. In *Agency in Archaeology*, M.-A. Dobres and J. Robb, eds. Pp. 40–50. London and New York: Routledge.
Wolf, Eric R.
  1997 *Europe and the People without History*. Berkeley, Los Angeles, London: University of California Press.
Woods, William I., and Joseph M. McCann
  1999 The Anthropogenic Origin and Persistence of Amazonian Dark Earths. *Yearbook, Conference of Latin Americanists Geographers* 25:7–14.
Woods, William I., Donald W. Meyer, and Joseph M. McCann
  2000 Black Earth Analysis: A Call for Cooperation. Paper presented at the Conference of Latin Americanists Geographers 2000, 6–8 January 2000, Austin, TX.
Yuval-Davis, N.
  1993 Gender and Nation. Ethnic and Racial Studies 16(4):621–632.
Zucchi, A.
  1985 Recent Evidence for Pre-Columbian Water Management Systems in the Western Llanos of Venezuela. In *Prehistoric Intensive Agriculture in the Tropics*, I. S. Farrington, ed. Pp. 167–180. BAR International Series 232. Oxford: British Archaeological Reports.

# INDEX

Page numbers *in italics* refer to illustrations.

## A

açaí (*Euterpeoleracea*), 62, 63, 92, 180
Acauan phase, 34, 38–39
Acrean Revolution, 171, 172
Acre region, 166, 169–170
   earthworks, 141–144, 146, 147, 163
   history of, 166–167, 170–172
   types of sites in, 149
   *see also* geoglyphs; Purus River region
Acuña, Cristóbal de, 108, 111–112
Açutuba (site), 20
ADE (Amazonian Dark Earth; *terra preta*), 18–19, 20, 123–126
   at Camutins M-17, 62
   composition of, 74
   in Tapajós River region, 26, 106, 113, 114, 123
   in Trombetas River region, 115, 120, 122, 123
   pH of, 18, 123, 125, 192
adornos (ceramic appliqués), 87, *87*, 120, 132
Africans (*quilombolas*), 185, 186, 190
age, 83, 86
agency, 35, 177
agriculture, 17, 71
   *see also* crops
agouti (motif), 131
Aldeia (site). *See* Santarém–Aldeia (site)
Aldeia Borary, 111
Amapá, 22
Amaral, Marcio, 119
Amazonas, 145–146
   *see also* Purus River region
Amazonia, 9, 11–12, 141
   Inca contact with, 158–159
Amazonian Dark Earth (ADE). *See* ADE (Amazonian Darth Earth)
Amazons, 105–106, 112, 129, 138
amulets, 129, *129*, 130
anacondas, 90
   *see also* snakes
Anajás River region, 46–47, 67, 71–72
Ananatuba phase, 33, 34, 38, 39
ancestors, 23, 69, 83–84, 110
animals, 186–187
   *see also specific animals*
Aparai, 90
Apurinã (Hypurinás; ethnic group), 168, 169, 174, 175
Aranha, Tenreiro, 167
Araona (ethnic group), 160–161, 171, 173, 174
Arauquinoid ceramics, 75
Arawak, 21, 22, 138, 154, 173
   architecture of, 157
   distribution of, *25*, 138, 173
   earthworks and, 163, 165, 174
   in Xingu River region, 165, 182
   Jesuit accounts of, 161
   landscape and, 188
   *see also* Baures (ethnic group)
archaeology, 30, 183
   indigenous, 184–185
   landscape, 77, 162
Archaic period, 12–13, 15
architecture, 32, 52–53
   defensive, 20, 164, 165

221

Armentía, Nicholas de, 161
Arripuna, 106
art, 79–80, 104, 181
    Chavín, 159
    geometric, 87–90, *88*
    representational, 80–81, *87,*
        87, 104
    rock, 12, *13*, 79
    see also figurines; iconography
Aruan phase, 33
astronomy, 12, 22
axes, 19, 86, 121, 128
ayahuasca, 174–175

## B

Balée, William, 124, 180
Balneário Quinauá (site), 150
bamboo, 124, 169
Barbosa Rodrigues, João de, 111, 114, 128
*barragens* (small dams), 48, 49, 51, 77n2
Barrancoid ceramics, 75, 120, 132
basalt, 126, 128
bat (motif), 87, 131
Bates, Henry, 111, 113–114
Baures (ethnic group), 73, 161, 165
beads, 31, 72, 83, 86, 106
beans, 17
Belterra Plateau, 116–117, 189
Bettendorf, João Felipe, 109–111
birds (motif), 87, 128, 131, 134
    see also owl (motif); vulture (motif)
bluffs (as site locations), 29, 113, 115, 120, 121
Boa Vista (site), 121, 132, 185–186
body parts (motifs), 87, 92–93, *93*
Bolívar, Gregorio, 173
Bolivia, 146, 147
    see also Llanos de Mojos
*bolsões*, *118*, 120, 134
BomFuturo (site), 118
bones (human), 82, 84, 94
    see also burial
Brivea, Domingo de, 108
burial, 26, 80–87, *95*

anthropomorphic vessels and, 24, 43, *87*, 92–98
    among Hypuriná, 169
    at Camutins, 63, 64, 84, *85*
    at Marajó Island, 19
    child, 83, 86
    during Formative, 19
    during Marajoara phase, 35
    iconography and, 84–86
    in mounds, 20, 83–84, *85*
    in urns, 22, 42, 43, 84, *85*
    lineage and, 83–84, 85
    Maracá, 22, 24
    mummies, 110
    secondary, 35, 42, 63, 82, 84, 169
burning, 181, 191–192

## C

cacao (*Theobroma cacao*), 124
caches, 21, *23*, 45
    see also bolsões
Cacoal (site), 71–72
Cacoal phase, 34, 71
caiman (motif), 81, 87, 95, 97, 131, 134
    ontangas, 101
calabashes, 21
calcium, 18, 123
Calçoene region, 22, *23*
Calusa, 70
*calzadas* (elevated roads), 75
Camará River region, 41, 42
*campos* (savannas), 30, 31, 64–65
Camutins (site), 26, 46–47, 49–56, 70, 71
    ceramic production area at, 60
    funerary vessels from 92–93
    M-1 (mound), 53, 56–60, *56,*
    M-17 (mound), 53, 60–64, 71, 84, *85*, 86
    population of, 52–53
    spatial analysis of, 53–56
    survey of, 49–51, 53
    tangas at, 100
Camutins River region, 46–48
canals, 161–162
Canamarys (ethnic group), 168

cannibalism, 82, 93
Capiranga (site), 117, *118*
capybara (*Hyrochoerus capybara*), 17
*caraipé* (bark), 21, 71, 120, 131
Carapanari (sites), 117
Carneiro, Robert, 66, 177–178
Carvajal, Francisco de, 29–30, 80, 105–106
Casa Velha sites, 43
Casinha (site), 67, 71
Castanheira (site), 38
*cauixi* (sponge spicules), 21, 120, 131
causeways, 50, 52, 73, 161–162
Cauyuaba (ethnic group), 73
Caverna da PedraPintada (site), 12, 15, 17, 21
Caviana (site), 38
cemeteries, 34, 83–84
  see also ancestors; burial
centralization
  gender and, 135–136
  hydraulic technology and, 59–60, 65–66
  see also elites
ceramics, 14–15, 34–35, 71, 72, 79–81
  as ethnic markers, 34
  basketry motifs, 132
  caches of, 21, *23*, 45, 120
  *caraipé*-tempered, 21, 71, 120, 131
  *cauixi*-tempered, 21, 120, 131
  hybrid images on, 81, 89, 95, 136
  Marajoara phase ceramics, 39–40, 41, 64
  Polychrome Tradition, 20, 96, 181
  Santarém phase, 97, 110, 131–136, *133*
  Tapajó, 96–97, 130–132
  workshops, 60, 64
  see also iconography; Incised Punctate Tradition; kilns
Cerritos (mounds in Uruguay), 151, 153
Chama, 98
Chandless, William, 167, 168–169, 172
charms, 129–130, *129*
Chavín culture, 159
chiefdoms, 70–71, 109, 163
  ascribed status in, 70, 83, 86
  trade and, 137–138
  see also chiefs
chiefs, 19, 70, 153
  see also chiefdoms
children, 83, 86, 114, 135
Chinésio (site), 147, *150*, 150
Choco, 104
chronology. *See* dating; seriation
circumscription, 66, 70
climate change, 18, 21
Clovis, 11
Codinach, Fidel, 171
Colombia, 22, 104, 130
  earthworks in, 26, 73, 75
Colorada Ranch (geoglyph), *145*
competition, 70
complexity (social), 65, 177, 182
  fishing and, 20, 25, 65–66
  Formative, 19–21
  see also chiefdoms; cultural ecology; polities (regional)
Coñori, 105, 111
copaiva, 166, 170
Coqueiral (site), 150
Coroca (site), 39
cosmology. see ideology; religion
Costa, Angione, 100
Costa Rica, 137
cremation, 42, 82
CRM projects, 121, 186
Croari (site), 38
crops, 16–17, 174
Crumley, Carole L., 154, 182
Cruz, Laureano de la, 109
cultural ecology, 65, 184
  perspective of, 22, 35–36, 142, 162, 177
cumaru, 170
Cuna (ethnic group), 104
Cunani River region, 22
*cupuaçu* (*Theobromagr and iflorum*), 124
currency, 128, 137
Curucurui (site), 117

**D**

dams, 48–49, 50–51, 162
   see also hydraulic technology
dating
   of Formiga phase, 39–40
   of M-1 (Camutins), 57, 59–60
   of M-17 (Camutins), 61–63
   of Marajoara phase, 35–36, 42
   of Marajó Island sites, 33, *34*, 35–36
   of Teso dos Bichos, 46
   see also seriation
deforestation, 189–190, *191*
   earthworks and, 27, 141, 172
Denevan, William, 162–163, 178–179
Desâna (ethnic group), 91
design. see iconography
*diretorias*, 167
ditches, 21–22, 146–148
   dating of, 148–149, 152
   discovery of in Acre, 141–144
   excavation of, 155–157
   function of, 157, 166
   in Llanos de Mojos, 163–166
   locations of, 145–146, *146*, 147
   photos of, *144*, *145*, *148*, *149*, *150*
   ring (geoglyphs), 21–22, 26–27, 145–150
   survey of, 143–144
   see also earthworks; hydraulic technology
diffusion, 15
disease, 107, 111, 174, 188
domesticates, 16–17, 174
   see also *specific plants*
drinking, 110, 111
   see also feasting
Dutch, 108

**E**

earthworks, 21–22, 26, 27, 72–73, 180
   chiefdoms and, 163
   dating of, 148–149, 152
   discovery of in Acre, 141–144
   Marajoara, 67
   photos of, *144*, *145*, *148*, *149*, *150*
   regional polities and, 72–77
   see also ditches, fields; geoglyphs; mounds
economy, 71, 138
   see also trade and exchange
ecotones, 17
Ecuador, 76
Eder, Francisco, 161, 165
effigy vessels, 132, 135, 136
electroresistivity survey, 44–45
elites, 56, 68–70, 128
   ceramics and, 40, 80–81
   genealogy and, 68–69, 81, 86
   iconography and, 135–136
enclosures. See ditches; geoglyphs
Erickson, Clark, 163, 165, 179
ethnoarchaeology, 125, 179, 182–183
ethnohistory, 29–30
Evans, Clifford
   at Camutins, 47, 49
   ceramic seriation of, 38–39, 41–42
   on Marajoara burials, 42, 86
   on Marajó Island, 33–36, 37
exotic goods. See prestige goods

**F**

face vessels (*vasos caretas*), 142, *143*
Faldas de Sangay (site), 26
Farabee, William Curtis, 42
Farias, João Barbosa de, 115
*farinha* (manioc flour), 16, 168, 189
Faro (site), 132
Fawcett, Percy, 169, 172–173
Fazenda Atlântica (geoglyph), 147, *148*, 155
Fazenda Colorada, 145, 147, 148–149, 150, 155, 156
Fazenda Iquiri (site), 150, 153
Fazenda Paraná (geoglyph), 147, *148*
Fazenda São Paulo (geoglyph), 148, 155
feasting, 20, 21, 70
feathers, 128, 134
fields, 74
   in Llanos de Mojos, 162–164, *164*
   raised, 26, 72–73, 74–75
figurines, 38, 80–81, 87, *87*, 102–104, *103*

at Camutins M-17, 63
broken, 103–104, 134
female, 102–104, *103*, 134–135
Konduri, 114
posture of, 110
Santarém phase, 132–134, *133*
stone, 114, 129–130, *129*
Tapajó, 97
*see also* hybrids; *muiraquitãs*
fish (motif), 128, 132
fish and fishing, 18, 26, 40, 126, 163
 in Trombetas River region
 on Marajó Island, 32–33
 poison and, 113–114
 regional polities and, 72–73
 snakes and, 90, 91, 92
 social complexity and, 20, 25, 65–66
 *see also* hydraulic technology
Flannery, Kent, 54–55
flooding, 32
floors, 58, *58*, 119
forests
 cultural, 124, 179–180
 national, 114, 121
"forest islands," 32, 180, 190
Formative period, 16–18, 181
Formiga (site), 39
Formiga phase, 33, 34, 39–40, 64
 ceramics, 39, 41, 42, 64
Fortaleza (site), 41, 42
fox (motif), 131
Franciscans, 108, 109, 160, 171, 175
 *see also* missionaries
Frei Luis (site), 42
Fritz, Father, 112
frog (motif), 94, 117, 131, 134, 138
 *muiraquitãs*, 117, 127–129, *127*
funerary practices. *See* burial

**G**
Galvan phase, 75
*gargalo* vessels, 132–134
Gavião (geoglyph), 148, *150*
Gavião cave, 13
Gê languages, 153
gender, 83, 85, 86, 96–97, 135–136
 *see also* figurines; women

genealogy: use by elites, 69, 81, 86
 *see also* kinship
Genipapo (sites), 118
geoglyphs (ring ditches), 22, 26–27, 144, 145–150, 157
 ceramics and, 155–157, 158, *158*
 dating of, 148–149, 152
 discovery of, 141–144
 excavation of, 154–157
 interpretation of, 174–175
 location of, 145–146, *146*, 147
 photos of, *144*, *145*, *148*, *149*, *150*
 spatial analysis of, 154, *154*
 survey of, 143–144
geology: of Marajó Island, 31
Goeldi Museum, 44, 121, 186
Goiapi River region, 41, *41*, 42–43
greenstone, 31, 126–129
 trade in, 19, 26, 31, 106, 136–138
Guajará mound, 61, 67
Guapindaia, Vera, 121–122, 134, 185
guaraná (*Paulliniacupana*; soapberry family), 17
Guari (site), 117
Guarita phase, 20
Guyanas, 26, 72, 74, 106

**H**
Hartt, Charles Frederick, 42, 97–98, 99
Hatahara (site), 20, 153
Heckenberger, Michael J., 21, 153–154, 182
Heriarte, Mauricio de, 110–111, 112
hierarchy. *See* stratification
Hilbert, Klaus, 120, 121, 185
Hilbert, Peter Paul, 115, 120, 121, 185
historical ecology, 9–10, 183
 *see also* landscape
houses, 52–53
hunter-gatherers, 11–13, 70
hybrids, 81, 89, 93, 104, 187
hydraulic technology, 19, 32, 48, 67
 at Camutins site, 49–51, 59–60

at Llanos de Mojos, 21, 65,
72–73, 163–164
ideology and, 66, 68–69
regional polities and, 72–73
*see also* lakes; fish and fishing;
ponds; wells
Hypurinás (Apurinã; ethnic group),
168, 169, 173, 174

## I
Ichipayo, 106
iconography, 87–90, 187–188
Chavín, 159
geometric, 87–90, 88, 171
of hybrids, 81, 89, 93, 104, 187
of tangas, 100–101
phallic, 102, 103
Santarém phase, 97, 132–136, 133
*see also specific motifs*
identity, 84–87, 88, 135–136
tangas and, 101–102
ideology, 80, 181, 188
gender, 96–97
hydraulic technology and, 66,
68–69
idols. *See* figurines
Igarapé dos Camutins, 46, 47,
50, 77*n*1
Ilha do Fogo (site), 39, 42
Inca, 158–160
Incised Punctate Tradition, 21, 26,
106, 113, 116, 131
distribution of, 113, 114,
122–123
in Tapajós River region, 106,
113, 122–123
ingá (*Inga edulis*; bean family), 17
intensification, 72–73
iron, 124

## J
Jacú (site), 118, 119
jade, 126–127, 137
*see also* nephrite
jaguar, 130, 131, 134, 159, 187
Japan, 15
Jesuits, 108–109, 110, 112,
161, 189

jewelry. *See* ornaments (personal)
Jivaro, 84
Joanes Painted vessels, 81, 92, 93, 96
Jomon ceramics, 15
Jurandir, Dalcídio, 91–92
Juruti (sites near), 21, 113

## K
kapok (*Ceibapentandra*), 124
Karajás, 52
kidnapping, 112, 170
kilns, 44
kinship, 56, 69, 86, 101
*see also* ancestors; genealogy
Konduri (ethnic group), 105–106,
111–113, 120–122
ceramics, 130–132
settlement organization of,
113, 114
subsistence, 112
Tapajó and, 136, 137–139
Kuikuru (ethnic group), 154,
177, 184

## L
labor, 51, 59–60, 68–71
geoglyphs and, 146–147
Labre, Antonio R. P., 170–171
labret, 38
Lago Grande (site), 20
lakes, 112, 114–115, 121
Lake Arari, 32
landscape, 10, 28, 64, 126
archaeology, 77, 162
domesticated, 73, 179
historical ecology and, 10, 183
in myth, 184, 186
regional polities and, 72–73
religion and, 23–24
transformation of, 23–24, 64,
72–76, 178–181
*see also* earthworks; geoglyphs;
hydraulic technology
language, 165, 173
Las Piedras (site), 160
Lathrap, Donald, 37, 104*n*1, 137,
159, 177–178
Lavras (site), 117, 118, 119

leadership, 71
  *see also* centralization;
    chiefs; elites
Leal (site), 71
lianas (*Paullinia pinnata*), 113–114
Limão Lake, 20
lime (industry), 14
lithics and lithic resources, 31, 121
  Archaic, 14, 15
  at Tapajó sites, 117, 123
  Paleoindian, 12, 13
lizard, 87, 128
Llanos (Venezuela), 26, 75
Llanos de Mojos (Bolivia), 19,
    161–163, 165, 181
  earthworks at, 21, 162–166, 181
  fields, 163–164, *164*
  hydraulic technology at, 21, 65,
    72–73, 163–164
  savanna burning at, 181, 191
Lobão (site), 142
looting, 14, 44, 46, 47, 56

**M**
Maciel Parente, Bento, 108
Magalis, Joanne, 96
magnesium, 18, 123
magnetic survey, 44–45
maize, 16, 21, 111
*maloca* (house form), 52–53
manatee (*Trichechusinunguis*), 17,
    112, 114
Manchineri (ethnic group), 175
Manentenery (ethnic group), 168
manganese, 18, 74, 123
mangroves, 14
Mangueiras phase, 33, 34, 38
manioc, 16–17, 55, 124, 126, 189
  among the Tapajó, 111
  graters, 121
  representations of, 159
  *see also farinha*
Maracá culture, 22, *24*, 97
Marajoara culture, 19, 30, 35, 36,
    40, 97
  collapse of, 22, 33, 71–72
  mounds, 34, 42, 67
  polities, 67–68
  settlement patterning, 41
  villages, 45–46, 52–53
Marajoara phase, 33, 34–36, 42,
    67–68
  Acauan phase and, 38
  ceramics, 39–40, 41, 64
  Formiga phase and, 39–40,
    41, 64
  houses, 52–53
  Incipient, 65–66
Marajó Island, 15, 19, 30–32, *31*,
    181, 188
  archaeology of, 30, 33–37
  during Formative, 17–18
  social complexity and, 19, 25
Marajó Project, 41–42
marriage, 110
Meadow croft Rockshelter (site), 11
megaliths, 22, *23*
Meggers, Betty J.
  at Camutins, 47, 49
  at Marajó Island, 33–36, 37
  ceramic seriation of, 38–39, 41
  cultural ecology and, 22, 36,
    37, 178
  on Marajoara burials, 42, 86
menstruation, 100, 101
Mesoamerica, 136–137
middens, 123, 125, 153
  *see also* shell middens
    (*sambaqui*)
migration, 11, 35, 174
minerals, 126–127
  *see also* greenstone; nephrite
missionaries, 108–109, 110, 112,
    160–161
missions, 111, 112, 160, 173–174
*mocambo* (fugitive slave community),
    112
Mojos (ethnic group), 73, 158–159,
    160
Mompós Depression (Colombia), 76
monkey, 29, 131
Monte Alegreregion, 12, *13*
  *see also* Caverna da Pedra Pintada
Monte Carmelo (site), 61, 86, 92
Monte Verde (site), 11
monumentality, 22, 69, 74, 157

*see also* earthworks; geoglyphs; mounds
mounds, 25–26, 27, 42, 150–154
    at Camutins site, 49–50, 51–52, 53–54
    ceremonial, 36, 42–43, 50, 53–54
    construction of, 19–20, 51–52, 59–60
    in Ecuador, 76
    in upper Xingu region, 21, 153–154
    in Uruguay, 151, 153
    Formiga phase, 39, 42
    in Goiapi River region, 42–44
    habitation, 36, 42, 50, 67
    M-1 (Camutins), 53, 56–60, 58
    M-17 (Camutins), 53, 60–64, 62, 71
    Marajoara, 34, 42, 67
    on Camutins River, 46–47, 46, 49–50
    spatial analysis of, 53–56
    symbolism of, 69
    types of, 36, 42, 49–50
    *see also* earthworks; monumentality
Mucajá (site), 39
*muiraquitãs*, 117, 127–129, *127*, 135
mummies, 110
Munduruku (ethnic group), 111
Muras (ethnic group), 168
musical instruments, 111, 130
Myers, Thomas, 37
myth, 26, 89, 165, 184
    snakes in, 90–92

**N**
Nazca lines, 143
nephrite, 72, 126, 129
Netto, Ladislau, 88, 100
Nhamundá–Trombetas River region, 114–115
    settlement patterning in, 106–107, 114–115, 120–122
    stone figurines from, 129–130
Nimuendajú, Curt, 113–115, 117–119
    surveys by, 21, 111, 120

Nordenskiöld, Erland, 138, 161–162
Noronha, Father, 114
nuts (Brazil; *Bertholletia excelsa*), 124, 166, 170, 174, 185

**O**
Omáguas, 52
Orellana, Francisco de, 29–30, 105–106
Oriximiná (site), 113, 115, 132, 138
ornaments (personal), *127*, 127–129, 135
    in burials, 83, 86
    pendants, 19, 117,138
    *see also* beads; greenstone; *muiraquitãs*; tangas (pubic covers)
owl (motif), 81, 93–94, 97, 134

**P**
Paço, Joaquim do, 166
Pacoval (site), 92, 94
Pacoval Incised vessels, 95, 96
Pacoval subphase, 45–46
Paititi, 73
Paiva, Paulo Marcelo, 139$n$1
Paleoindians, 11–13
palms, 17, 32, 36, 124
    domestication of, 13–14, 17
    spirits, 175
    use of, 32, 74, 174
    *see also* açaí (*Euterpeoleracea*)
Panoan, 98, 173
papaya (*Carica papaya*), 117, 124, 174
Parapará (site), 42
Paredão phase, 20
Parente, Bento Maciel, 108
Pärssinen, Martti, 143, 154–155, 160
Pedra Pintada. *See* Caverna da Pedra Pintada (site)
pendants. *See* ornaments (personal)
Penna, Ferreira, 44, 100
peppers (*Capsicum* spp.), 17
Pequiá cave, 13
perspectivism, 187
phosphorus, 18, 123, 126
phytolith analysis, 17

Pindobal 1 (site), 117, 118
pineapple, 17
pipes, 134–135
Piquiatuba (site), *127*
pirarucu (*Arapaima gigas*),17, 170
place names, 171
Plafker, George, 162
plants, 13, 170, 179–180
   see also specific plants
platforms, 50, 52, 63, 76
Pleistocene,11–12
Pocó (ceramic type), 120–121
poison, 105, 113–114, 126, 169
political economy, 71, 138
polities (regional), 23, 29–30, 33, 123
   Classic Marajoara, 67–68
   development of, 24, 25, 29, 67–68, 72
   earthworks and, 72–77
   trade and, 19, 23, 123
Polychrome Tradition, 20, 96, 181
polygamy, 110
ponds, 32, 49, 50, 51
population, 17, 18, 124, 188–189
   at Camutins, 67–68
   decline, 107, 112, 188
   geoglyphs and, 157
   of Marajoara villages, 45–46, 52–53
Porto (site), 21, 117, *118*, 119–120
Port site. *See* Porto (site)
Portuguese, 29, 107–109
post holes, 45, 155
Posto Novo (site), 118
pottery. *See* ceramics
precipitation, 32
pregnancy (motif), 97, 102, 134
preservation (archaeological), 13, 16
prestige goods, 23, 40, 128, 137
   in burials, 83, 86
   see also greenstone
puberty ritual, 98, 100, 101–102
   see also tangas
Puru-purus (ethnic group), 168
Purus River region (upper), 166–167, 168, 170–172
   Arawaks and, 173
   burial practices in, 169

Q
Quechua language, 171, 173
*quilombolas* (African descendants), 185, 186, 190
Quinari Tradition, 142,

R
raiding, 107
Ramal do Capatará (site), 149, 155–157
ranching, 27, 92, 172, 189
   effects of, 49, 50, 141, 191
rank, 69–70, 83, 86
rank-size rule, 54–55
Ranzi, Alceu, 142–144, 157
rattles, 102
Reichel-Dolmatoff, Gerardo, 104
religion, 34, 79, 97, 157
   landscape and, 23–24
   see also ideology
reproduction, 90
rice, 112
ritual, 23, 82, 157
   elites and, 69, 81, 188
   figurines and, 104, 134
   integration, 70, 75
   puberty, 98, 100, 101–102
   termination, 20, 120, 153
   see also burial; feasting
roads, 30, 75, 160, 180
   geoglyphs and, 145, 147–148, 153, 157
   in Llanos de Mojos, 73
   upper Xingu, 21, 154
rock art, 12, *13*, 79
Rodrigues Ferreira, Alexandre, 91
Rondônia, 145
Roosevelt, Anna, 14, 15, 36–37, 100
   on Marajoara collapse, 71
   at Teso dos Bichos, 43–45
   see also Porto (site)
rubber (*Heveabrasiliensis*), 166–167, 169–172

S
Saint John (province), 105, 106, 112
Salitre (site), 42
Sangay (site), 76

Santa Izabel (geoglyph), 147, *149*
Santarém (city), 111
Santarém–Aldeia (site), 21, 113, 117, 119, 120
   ADE and, 123
   workshop at, 128
Santarém phase, 119
   ceramics, 97, 110, 131–136, *133*
   figurines, 132–134
   iconography, 135–136
   pipes, 134–135
   *see also* Porto (site); *and under* Tapajó culture
Santa Teresinha (geoglyph), 147, *149*
Santiago (site), 76
Santos-Granero, Fernando, 186, 188
São Domingos (sites), 117
São Leão (site), 39
Saparará (site), 67
Sapucuá Lake sites, 115, 132
sarsaparilla, 166, 170
Saunaluoma, Sanna, 155, 164–165
savannas (*campos*), 30, 31–32, 64–65
scorpion (motif), 94, 95, 97
sedentism, 14, 125
seriation, 33, 178
   *see also* dating
settlement patterning, 37, 154, 182
   at Camutins, 53–56
   hierarchical, 36, 53–56, 76,
   in Tapajós River region, 106–107, 113, 114–119, 181–182
   Marajoara, 41
   on bluffs, 29, 113, 115, 120, 121
Seu Chiquinho (geoglyph), *144*
Severino Calazans (site), 148
shamans, 19, 23, 97, 104, 130, 187
shellfish, 14, 15, 16
shell middens (*sambaqui*), 14, 15, 33
Silva Salgado, Serafim da, 167
Sinú culture, 75
Sipó Incised ceramics, 39
slaves, 86, 107, 108, 166
   African, 112, 185
   Tapajó and, 108–109
snakes, 26, 90–92, 101, 128, 134

motif, 87, 89–90, *89*, 94, *95*, 97
sodium, 124
soil, 30–31, 125, 126, 192
   *terramulata*, 117, 122, 125
   *see also* ADE (Amazonian Dark Earth; *terra preta*)
soul, 82, 91, 94
Souza, Manoel de, 112
Spaniards, 29–30, 105–106, 166
   Inca and, 158–160
specialization (ceramic), 35
spindle whorls, 39, 72
"staple finance," 68
stratification, 23, 34, 66, 136
stratigraphy (mound), 56–59, *58*, 61–63, *62*
status. *See* rank
Steere, J., 44
stools, 81, 98
storage, 71
stoves, 44–45, 120
subsistence, 11–17, 18
   *see also* fish and fishing
Suriname, 75
surplus, 68, 70, 71
sweet potatoes, 16–17
symbolism, 24, 69, 81, 137
   elite use of, 65–66
   *see also* iconography; snakes

T
Takana speakers, 160, 161, 171, 173–174
tangas (pubic covers), 63, 72, 88, *95*, 99, 97–102
   burial urns and, 42, 44, 100
   puberty ritual and, 98, 100, 101–102
   use wear on, 98, 100
Tomorona (ethnic group), 161, 173
*tapagem* (wall dam), 48, 49, 77n2
Tapajó culture, 20–21, 106, 108–111
   burial practices of, 82, 97, 120
   ceramics, 96–97, 130–132
   Konduri and, 136, 137–138

trade and, 106, 127, 128, 137–138
  *see also* Porto (site); Santarém phase
Tapajós River region, 18, 19, 20, 26, 123
  settlement patterning in, 106–107, 113, 114–119
  survey of, 114, 116–117
Taperinha (site), 14, 15
tapioca, 16
Taquara-Itararé Tradition, 152
Teixeira, Pedro, 108, 109
*terra firme* (interfluvial land), 20, 21, 22, 177, 179, 180
*terra mulata*, 117, 122, 125
*terra preta* (dark earth). *See* ADE (Amazonian Darth Earth)
Terra Preta do Jacú (site), *118*, 119
Terra Santa (site), 132
territoriality, 12, 56, 66, 69, 79, 84
Teso dos Bichos (site), 41, 42, 43–45, 52, 67
Teso subphase, 45–46, 71
*timbó* (*Paullinia pinnata*; soapberry family), 113–114
Tinamostón, 106
Tocantins, 92
totems, 81
trade and exchange, 31, 68, 71, 123
  after contact, 107–108
  regional polities and, 19, 23, 123, 136–138
  with Andes, 27, 159
  *see also* greenstone
trampling, 50
transportation, 48, 59
  *see also* roads
tribute, 66, 71
trickster (motif), 94
Trombetas River region, 20, 26, 111, 112, 123
  settlement patterning in, 107, 113–115
  *see also* Konduri (ethnic group)
Tukano (ethnic group), 90
*Tuluperê* (mythic snake), 90–91

Tumichucua (site), 164
Tupi (ethnic group), 17, 22–23
Tupi-Guarani people, 17
turtles, 15, 29, 30, 191
  motif, 87, 128

**U**
Uaboi (ethnic group), 112, 114
upper Purus River region. *See* Purus River region
upper Xingu River region. *See* Xingu River region
Urbano, Manoel, 167–168
urn burials, 22, 42, 43, *43*, 84, *85*
*uruá* (shellfish), 16
Urucurus (ethnic group), 109, 111

**V**
vagina (motif), 92, *93*, 95, 96, 97, 101, 102, 135
*várzea* (floodplain), 177–178, 179
  cultural ecology and, 22, 65
  social complexity and, 19, 20
Vega, Garcilaso de la, 158–159
Venezuela
  ceramic styles of, 131, 138, 179, 181
  earthworks in, 26, 72, 75
Vieira, Antonio, 109
Vila Americana (site), 117
Vila Franca sites, 111, 113, 115, 123
villages, 17, 19, 20, 52
  at Teso dos Bichos, 45–46
  circular, 20, 157
  population of, 45–46, 52–53
  *see also* settlement patterning
violence, 107, 108, 110
  *see also* warfare
Virtanen, Pirjo, 174–175
Vista Alegre (site), 71
vulture (motif), 95, 97, 131

**W**
Walker, John H., 163–164, 165
walls. *See* earthworks
Wanâna, 91

warfare, 66, 71, 107
water, 32, 47–48, 147
   *see also* hydraulic technology
Wayana, 90
wells, 118–119
wives, 19
womb (motif), 92, 93, 94, 95, 96
women, 86, 88, 114, 135
   representations of, 97–98, 135–136
   *see also* figurines; gender; tangas (pubic covers)
workshops, 60, 64, 128

X, Y, Z
Xingu River region (upper), 21, 153–154, 165, 182
*zanjas circundantes*. *See* ditches
zinc, 18, 74, 123
Zinha (site), *118*

# About the Author

An internationally renowned Amazonian archaeology specialist, Dr. Denise Schaan is best known for her innovative research on Marajó Island, at the mouth of the Amazon river, and more recently for her leading role in the western Amazonia geoglyph research, which have attracted worldwide attention.

Schaan completed a master's program in archaeology in 1996 in Brazil, and earned her Ph.D. at the University of Pittsburgh in 2004. In 2005, she joined the faculty of the Department of Anthropology of the Federal State University of Pará, in northern Brazil, where in 2010, together with other colleagues, she created the first Four Field Anthropology Graduate Program in the country.

Schaan has published 44 scientific articles and book chapters focusing on Amazonian archaeology, two monographic books and two collections of essays on the archaeology and ceramic iconography of the Marajoara culture, and two organized books on the Acrean geoglyphs. Her comparative landscape approach to Amazonian archaeology is innovative because, with few exceptions, archaeologists working in the area tend to focus their attention on single sites or small groups of sites. Carrying out regional-scale fieldwork in different locations of the Amazon River basin (upper and lower), Schaan has been able to get firsthand data on sociocultural developments in the basin relating to impressive landscape modifications and also to compare indigenous Amazonian strategies in different times and places.